Platform Engineering for Architects

Crafting modern platforms as a product

Max Körbächer

Andreas Grabner

Hilliary Lipsig

Platform Engineering for Architects

Group Product Manager: Preet Ahuja

Publishing Product Manager: Suwarna Patil

Book Project Manager: Uma Devi

Senior Editor: Adrija Mitra

Technical Editor: Nithik Cheruvakodan

Copy Editor: Safis Editing

Proofreader: Adrija Mitra

Indexer: Tejal Soni

Production Designer: Alishon Mendonca

Senior DevRel Marketing Executive: Rohan Dobhal

First published: October 2024

Production reference: 2050126

Published by Packt Publishing Ltd.
Grosvenor House
11 St Paul's Square
Birmingham
B3 1RB, UK

ISBN 978-1-83620-359-9

www.packtpub.com

To my wife, best friend, partner in crime, and North Star, Lesya, for being my loving partner and biggest supporter, who always has my back and calls my creativity.

- Max Körbächer

To my parents, who allowed and supported me to follow a different career path than they initially had in mind. Thanks for being the best role models one could wish for.

- Andreas Grabner

To my husband, Scott, without whom I'd never have become an engineer. To my parents, Annette and Calvin, who have loved me and believed in me – your encouragement guides me. To my parents-in-law, Joe and Mary Ann, who have been there every step, and to my kids, who support my dreams and still think I'm fun to hang around. Thank you for everything, all of you.

- Hilliary Lipsig

Contributors

About the authors

Max Körbächer is a technology advisor and platform architect who focuses on utilizing cloud-native technologies and open source to simplify the challenges of complex systems. He is the founder and managing director of Liquid Reply, a cloud-native engineering and consulting company. His work history includes roles as an enterprise architect in the media and power utility industry and as a demand manager planning medium and large IT projects. Max is also the founder and, currently, co-chair of the CNCF Environmental Sustainability Technical Advisory Group, CNCF Ambassador, Linux Foundation Europe Advisory Board member, and initiator and organizer of the Kubernetes Community Days Munich and Ukraine.

I want to thank my family – especially my wife, Lesya; my best writing supporter, Felix, who stayed for all the late-night writing sessions; and all the open source community people for discussing ideas, solutions, and the future of cloud native.

Andreas Grabner is a technical advocate for making distributed systems observable and making automated data-driven decisions across the software development lifecycle. In his capacity as a CNCF ambassador and a DevRel at Dynatrace, he connects and educates global software engineering communities on building and continuously validating digital services for resiliency, high availability, and security.

Since his early days, he has been passionate about software quality and performance engineering as it results in building excellent digital products. Andi uses his advocacy platforms to share best practices on topics such as observability, progressive delivery, DevOps, site reliability engineering, platform engineering, and digital business operations!

I want to thank my wife, Gabi, for being supportive throughout the whole process, especially while writing during our vacation. Also, a big thanks to my employer and all the global tech communities that supported me along the way!

Hilliary Lipsig is an autodidact and start-up veteran who has frequently learned and applied technologies to get a job done. She's had her hand in every part of the application delivery process, honing her skills originally as a quality engineer. Hilliary is an IT polyglot, able to talk the lingo of both the Operations and Development teams. She's currently a Principal Site Reliability Engineer at Red Hat Inc., working on Kubernetes-based platforms. She's passionate about GitOps, continuous integration, scalable processes, consistency in tooling, and good developer documentation. Her open source activities include contributions to the CNCF Glossary and she's a member of the Code of Conduct Committee for Kubernetes.

A big thank you to my family, friends, and colleagues who supported me, let me spitball ideas at them, and even volunteered to proofread for me. Your belief in me helped me believe in me.

About the reviewers

Peter Portante has been a Red Hat, Inc. employee since late 2011, working as a senior principal software engineer in the Performance and Scale team for the first 12 years and as an Enterprise Account Solution Architect since the spring of 2023. Prior to Red Hat, he worked at HP/Compaq/Digital on HP ePrintCenter, Tabblo, TruClusters, gWLM, POSIX Threads (two-level scheduling in Tru64 Unix and OpenVMS), and the OpenVMS Print Symbiont.

Fábio Falcão has been a technology professional for over 20 years, with solid training as a SysAdmin and in-depth knowledge of DevOps methodologies and tools. Having graduated in computer science and being passionate about technology and how it helps improve people's lives, he decided to pursue this career as a child and, today, really loves what he does, helping teams achieve their goals. From Brazil and living in Portugal since 2022, he joined IBM to work with the data lineage team, in addition to supporting several other products linked to artificial intelligence.

I would first like to thank my parents (in memoriam) for all their dedication and encouragement throughout my life. I would also like to thank my wife and children, who support me when I don't have enough time for them, and my friends, who always help me grow professionally and personally with their advice and chats.

Lukasz Bielinski is a seasoned senior DevOps leader with 15 years of experience in IT, specializing in containerization and digital transformation for large-scale systems across banking, telecommunications, pharmaceuticals, and software industries. He has successfully led and executed numerous Kubernetes projects, significantly enhancing operational efficiency and system reliability. Lukasz co-founded and formalized a start-up where he played a crucial role in introducing Kubernetes and OpenShift into organizations, driving successful digital transformations. His technical acumen spans a wide range of DevOps tools and practices, making him a trusted advisor and leader in the field.

I would like to express my gratitude to my family for their patience and support as I dedicated time to reviewing this book. Special thanks to the authors for entrusting me with this important task and to my colleagues for their ongoing encouragement. I'm grateful to be part of a community that values continuous learning and collaboration.

Thomas Schuetz is a cloud native architect with a keen interest in cloud-native application delivery. He teaches at an Austrian University of Applied Sciences, focusing on cloud-native technologies, where he shares his industry insights and practical experiences with students. Involved in the cloud-native community as a CNCF Ambassador, Thomas is enthusiastic about open source projects, contributing as a Keptn GC member. His approach to software delivery and troubleshooting combines practical know-how with a passion for education, aiming to make cloud-native technologies accessible to all.

I am deeply honored to have been invited to review this book by thought leaders in the cloud-native community. Contributing to this work has been both satisfying and rewarding, and I thoroughly enjoyed reading it. I would also like to express my gratitude to the entire cloud-native community for their continuous innovation and for making advancements such as platform engineering possible.

Biswajyoti Chowdhury has over 13 years of experience in cloud-native platform engineering, has led the development of multiple internal development platform initiatives, and has successfully migrated several large enterprises to the cloud and established their development platforms. He has worked with prominent technology consulting firms such as Accenture, Wipro, and Cognizant. Currently, he serves as an architect at Financial Software & Systems, where he is focused on building a payments-centric development platform. Biswajyoti is passionate about open source, cloud-native infrastructure, and tools that enhance developer productivity. Biswajyoti holds industry certifications such as GCP PCA, ACE, and is a Red Hat Accredited Professional in Cloud-Native development.

I would like to thank my family, my mentor, and my colleagues, who have put up with me and shaped me into who I am today.

Table of Contents

Part 2 – Designing and Crafting Platforms

4

Architecting the Platform Core – Kubernetes as a Unified Layer   117

5

Integration, Delivery, and Deployment – Automation is Ubiquitous   157

6

Build for Developers and Their Self-Service   199

Part 3 – Platforms as a Product Best Practices

7

10

Crafting Platform Products for the Future   317

Preface

Hey, welcome to *Platform Engineering for Architects*! Platform engineering is the practice of creating environments that can build, test, validate, deploy, and operate software in a secure and cost-efficient way. Platform engineering is about automation, and about enabling the platform's users, developers, and operations to focus on value creation. A platform, often defined as an **Internal Development Platform** or **Internal Developer Portal** (**IDP**), abstracts away the complexity of the underlying infrastructure and all the moving components required to support the software life cycle from its setup until it goes live in production. But platform engineering is more than just technologies that have to play well together. It requires an open mindset and a holistic approach to define the purpose of the platform, along with following principles in the decision-making process and fostering a culture of change and innovation.

In *Platform Engineering for Architects*, we will engage you in building up a product mindset to build a solution that ages and matures with time but stays young at heart. Step by step, we will create our strategic direction and define our target architecture for the platform. This will become a living artifact for you and everyone working with you. Throughout the book, we will cover the four different pillars of a platform and the relevant decisions you have to take: the infrastructure representation by Kubernetes, the automation, the self-service capabilities, and the built-in observability and security. By the end, you will be equipped with tools to handle costs including actual infrastructure costs and technical debts. If both are managed well, they can even become your best allies in overcoming organizational obstacles and politics.

For every aspect and topic we write about in this book, we also provide additional sources that cover full technical details. Our goal is to provide you with a framework of references and an approach for defining platform architectures that can mature over time without being tied to a specific technology or version. We are well aware that the only constant in life is change, and this holds true for platform engineering. That's why we also encourage you as a platform engineer and architect to keep up to date with those changes and bring the product mindset to your users and stakeholders.

Who this book is for

The book is for platform engineers and architects, DevOps engineers, and cloud architects who want to transform their way of implementing cloud-native platforms to use a platform as a product.

This book is also for IT leaders, decision-makers, and IT strategists who are searching for new approaches to improve their systems landscape and software delivery, covered by a holistic approach that goes beyond the simple "*you built it, you run it.*"

What this book covers

Chapter 1, Platform Engineering and the Art of Crafting Platforms, provides an introduction to platforms and IDPs. It covers the relevance of a product mindset and the ambition to build a system desired by users.

Chapter 2, Understanding Platform Architecture to Build Platforms as a Product, will guide you through all the relevant groundwork and approaches to creating your platform architecture. You will discover the value of the platform as a product, the first implementation of the thinnest viable platform, and how to observe and measure its success and adoption.

Chapter 3, Building the Foundation for Supporting Platform Capabilities, will walk you through the mandatory steps and processes of defining a solid foundation of a platform that can grow from an initial set of features toward key enterprise-supporting platform capabilities.

Chapter 4, Architecting the Platform Core – Kubernetes as a Unified Layer, provides insights into what makes Kubernetes the preferred platform for platform engineers. You will learn about the core integrations and relevant decisions we have to make before we can focus on extra enhancements.

Chapter 5, Integration, Delivery, and Deployment – Automation is Ubiquitous, provides you with a stable understanding of the complexity around building, deploying, testing, validating, securing, operating, releasing, and scaling software and how we can centralize and automate this experience with self-service capabilities.

Chapter 6, Build for Developers and Their Self-Service, reviews concepts around IDP integrations and shares best practices for building resilient, flexible, and user-oriented platforms.

Chapter 7, Building Secure and Compliant Products, elaborates on security standards frameworks, and trends; how to leverage the software bill of materials; and defining the right actions to secure the platform without limiting your capabilities. Furthermore, we show you how to ensure the app delivery process will provide hardened and secure software/container packages and how to use policy engine technologies.

Chapter 8, Cost Management and Best Practices, explains the concept of cost-increasing elements of a platform and how to optimize those costs. You will learn about tagging strategies, general cost optimization scenarios, how to use observability to identify optimization potential, and best practices to put them into practice.

Chapter 9, Choosing Technical Debt to Unbreak Platforms, provides you with tools, frameworks, and methods to actively manage your technical debts. Like with costs, technical debts can grow and will have a negative impact on your platform if untreated.

Chapter 10, Crafting Platform Products for the Future, emphasizes the imperative of change and our role as platform engineers in fostering change in a controlled way, balancing reliability and innovation.

To get the most out of this book

You should understand the basics of cloud computing, Kubernetes, the ideas around platform engineering, and how to define those architecture-wise.

Software covered in the book
Kubernetes
Backstage
CI/CD solutions such as GitHub Actions
Keptn
Argo CD
Crossplane
Prometheus
OpenTelemetry
Harbor
OpenFeature
Renovate Bot

Download the example code files

You can download the example code files for this book from GitHub at `https://github.com/PacktPublishing/Platform-Engineering-for-Architects`. If there's an update to the code, it will be updated in the GitHub repository.

We also have other code bundles from our rich catalog of books and videos available at `https://github.com/PacktPublishing/`. Check them out!

Conventions used

There are a number of text conventions used throughout this book.

`Code in text`: Indicates code words in text, database table names, folder names, filenames, file extensions, pathnames, dummy URLs, user input, and Twitter/X handles. Here is an example: "The directory path is nested within the chart, and the use of this directory called `crds/` allows Helm to pause while the CRDs are added to a cluster before continuing with the chart execution."

A block of code is set as follows:

```
apiVersion: apiextensions.k8s.io/v1beta1
kind: CustomResourceDefinition
metadata:
  name: crontabs.stable.example.com

spec:
  group: stable.example.com
  version: v1
  scope: Namespaced
  names:
    plural: crontabs
    singular: crontab
    kind: CronTab
    shortNames:
    - ct
```

Any command-line input or output is written as follows:

```
$ kubectl label nodes platform-worker2 reserved=reserved
node/platform-worker2 labeled
```

Bold: Indicates a new term, an important word, or words that you see onscreen. For instance, words in menus or dialog boxes appear in **bold**. Here is an example: "These platform logs should also, as much as possible, be clear of any **Personally Identifiable Information (PII)**."

> **Tips or important notes**
> Appear like this.

Get in touch

Feedback from our readers is always welcome.

General feedback: If you have questions about any aspect of this book, email us at customercare@packtpub.com and mention the book title in the subject of your message.

Errata: Although we have taken every care to ensure the accuracy of our content, mistakes do happen. If you have found a mistake in this book, we would be grateful if you would report this to us. Please visit www.packtpub.com/support/errata and fill in the form.

Piracy: If you come across any illegal copies of our works in any form on the internet, we would be grateful if you would provide us with the location address or website name. Please contact us at copyright@packt.com with a link to the material.

If you are interested in becoming an author: If there is a topic that you have expertise in and you are interested in either writing or contributing to a book, please visit authors.packtpub.com. Share Your Thoughts

Share Your Thoughts

Now you've finished Platform Engineering for Architects, we'd love to hear your thoughts! Scan the QR code below to go straight to the Amazon review page for this book and share your feedback.

https://packt.link/r/1836203594

Your review is important to us and the tech community and will help us make sure we're delivering excellent quality content.

Stay Sharp in Cloud and DevOps — Join 44,000+ Subscribers of CloudPro

CloudPro is a weekly newsletter for cloud professionals who want to stay current on the fast-evolving world of cloud computing, DevOps, and infrastructure engineering.

Every issue delivers focused, high-signal content on topics like:

- AWS, GCP & multi-cloud architecture
- Containers, Kubernetes & orchestration
- **Infrastructure as Code (IaC)** with Terraform, Pulumi, etc.
- Platform engineering & automation workflows
- Observability, performance tuning, and reliability best practices

Whether you're a cloud engineer, SRE, DevOps practitioner, or platform lead, CloudPro helps you stay on top of what matters, without the noise.

Scan the QR code to join for free and get weekly insights straight to your inbox:

https://packt.link/cloudpro

Free Benefits with Your Book

This book comes with free benefits to support your learning. Activate them now for instant access (see the "*How to Unlock*" section for instructions).

Here's a quick overview of what you can instantly unlock with your purchase:

PDF and ePub Copies

Next-Gen Web-Based Reader

Access a DRM-free PDF copy of this book to read anywhere, on any device.

Use a DRM-free ePub version with your favorite e-reader.

Multi-device progress sync: Pick up where you left off, on any device.

Highlighting and notetaking: Capture ideas and turn reading into lasting knowledge.

Bookmarking: Save and revisit key sections whenever you need them.

Dark mode: Reduce eye strain by switching to dark or sepia themes

How to Unlock

Scan the QR code (or go to packtpub.com/unlock). Search for this book by name, confirm the edition, and then follow the steps on the page.

UNLOCK NOW

Note: Keep your invoice handy. Purchases made directly from Packt don't require one

Part 1 –
An Introduction to Platform
Engineering and Architecture

In the first part, you will learn more about the foundation of platform engineering and platform architecture. We will sharpen your understanding of a platform product mindset, showing you that platform engineering is more than just building systems. Then, we will guide you through the process of creating a platform architecture, from defining the purpose of a platform to the thinnest viable platform, as well as how to measure a platform's success and acceptance. In *Chapter 3*, we will focus on designing the foundation of a platform to enhance the user experience and avoid technical complexity.

This part has the following chapters:

- *Chapter 1, Platform Engineering and the Art of Crafting Platforms*
- *Chapter 2, Understanding Platform Architecture to Build Platforms as a Product*
- *Chapter 3, Building the Foundation for Supporting Platform Capabilities*

1

Platform Engineering and the Art of Crafting Platforms

In this first chapter, we will learn how to identify when our organization is in the right state to plan a platform. For this, we will clarify why platforms have become such a relevant topic, how a product mindset fits into this, and what the checkpoints are to find out whether we are ready for a platform or not. We will learn about the platform differences and which platform types are most commonly built.

Next, we will delve into the three core elements of a platform: the pervasive cloud, the developer experience, and the main attributes of a platform. Overall, we will see recurring elements of cloud-native engineering. This leads us to the question of whether we really need yet another abstraction layer. We will also consider whether a platform will help us to overcome the problem of a high cognitive load caused by overengineered complex systems and development processes or just end up being yet another layer. We will reflect on some of those layers to find an answer for ourselves.

Finally, we will go into aspects that go beyond the technology and the implementation of platforms. It is crucial to understand the sociotechnical aspects and put the human, our actual stakeholders, at the center. This allows us to define a better platform product and find approaches for a close collaboration.

In this chapter, we're going to cover the following main topics:

- The demand for platforms as a product
- Implementing developer- and product-focused solutions
- Do we need yet another abstraction layer?
- Sociotechnical aspects

Free Benefits with Your Book

Your purchase includes a free PDF copy of this book along with other exclusive benefits. Check the *Free Benefits with Your Book* section in the Preface to unlock them instantly and maximize your learning experience.

The demand for platforms as a product

In the cloud-native environment, hardly any other topic has built up such a myth in recent years as the term *platform* and the associated role of the platform engineer. As with the introduction of the first usable CI/CD pipelines, this gold rush led to rapid adaptation, often without sense or reason. Now that we have arrived in the valley of knowledge, we can deal extensively with the question: do you need a platform, and if so, how do you design and implement it to ensure that it lasts into the future?

To answer this question, we should first look at what constitutes such a platform. A platform is the combination of different capabilities that are required to master traditional and cloud-native environments so that it supports the end user in the development, delivery, and operation of an application. Platforms can be an enabler to turn non-cloud native infrastructures into valuable resources. However, most computing platforms today provide some sort of API that can be used to automate the deployment and instrumentation of the available resources and build the foundation of a platform. Platforms provide consistency across any kind of resources for the end users and grant access to its capabilities via a self-service API, templates, CLI, or other solutions. The following example also highlights that a platform is composed of many components:

Figure 1.1: Example of a platform/IDP

We see usually the topic of platform appears in the context of cloud-native, but why is that so? Cloud-native technologies enable organizations to build and run scalable applications in public, private, and hybrid clouds. This approach is best illustrated by functionalities such as containers, standardized service provisioning, immutable infrastructure, and declarative APIs. Such functionalities realize loosely coupled systems that are resilient, manageable, and observable. These enable developers to make frequent changes with minimal effort. In short, a platform is an enabler for cloud-native computing and uses its tooling to instrumentalize it.

Companies and developers benefit from platforms in an equal manner

The experience of a software engineer on a cloud-native platform differs from developing software natively toward a cloud provider. Building systems focusing on one **Cloud Service Provider** (**CSP**) will bind you to the logic of that closed ecosystem. You will surely have a similar effect when you build on cloud-native platforms due to the fact that those are often Kubernetes-centric, utilizing the heavy unification of integrations toward the Kubernetes API. However, the catch is that cloud-native platforms deliver the same experience without you recognizing the underlying infrastructure. As most companies have at least two to three cloud or cloud-like service providers and already have difficulties in adapting those, a cloud-native platform is a game changer *[1]*. Developing software on a cloud-native platform changes the mindset and architecture. However, without adopting that mindset, the chance of failure is high.

However, there are more aspects to consider for a platform than just unified infrastructure management. Platforms have to be made for a purpose. The common definition of whom platforms are built for, and who the stakeholders for platform engineers are, states that those are exclusively developers. These definitions fall short of mentioning that a whole organization, operational teams, and other specialist teams also benefit from a platform. A platform provides software engineers with a simple access point to build, test, deploy, release, and operate their software. It provides deep insights into the usage and allows the caretaker and administrators to maintain the infrastructure, platform, and integrations fearlessly. To translate this into business terminology, a platform can provide a faster time to market, with more flexibility to change and adjust its components, while keeping reliability and robustness high.

What does this mean for a company now? Due to the shortage of IT professionals on the market, the fast pace of changes in IT, and the overload of training teams for cloud technologies and providers, a platform introduces the right breakpoints for competencies. We need these breakpoints to declutter the trend of putting multiple disciplines into a single role such as DevOps. Also, platform engineers utilizing DevOps methodologies are not DevOps. We actively need to protect this role from repeating the mistakes made with the DevOps role and stay sharp in its definition. Platform engineers integrate experts' provided capabilities, simplify their usage of those capabilities through their platform for developers, and enable self-service for the engineers. However, no developer will need to become an expert in multiple topics such as security, observability, infrastructure configuration and automation, and so on. This is in contrast to a common picture of DevOps, who need to become experts with anything that is required within their silo for their application to keep them alive. We will need DevOps in the future for the advanced handling of applications, but we must make their lives easier, too.

The platform provides an integration layer for the bottom-up capabilities that require special knowledge such as security, databases, or even the deployment of VMs or bare-metal servers, as well as top-down usage by developers and DevOps. As visualized in the following figure, the platform engineering team is responsible for providing this layer.

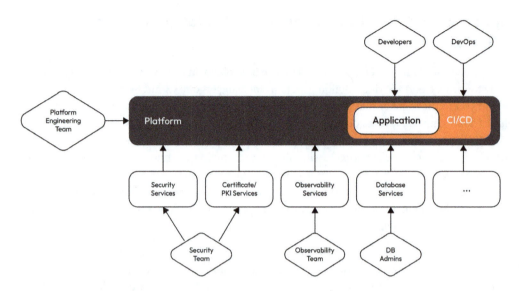

Figure 1.2: Capabilities and responsibilities in a platform-driven organization

Of course, this also means that another team of experts must be trained and educated. However, a comparably small team of platform engineers can usually build and run huge environments. A platform ideally reduces the cognitive load for any other team within the company and lets them focus on their core value again by simplifying the machinery around the development process. This platform helps to reduce the stress and improve transparency. Companies from all over the world frequently share their experience with platforms and platform teams, and the typical tenor is on how they solved problems they couldn't tackle before, or how much this has improved the quality of their products and services.

These platforms are often called **Internal Developer Platforms (IDPs)** because they are usually built for an enterprise's internal development team. Throughout this book, we will use the terms platform, IDP, platform product, and cloud-native platform interchangeably. However, we'll first highlight certain aspects just a little bit more:

- **Platform**: General term for the cross-cutting layer of technology that allows unification of services for developers.

- **IDP**: Emphasizing the aspect of developer, **Software Development Life Cycle (SDLC)**, and tools needed to develop software

- **Platform product or platform as a product**: Highlighting the dedicated team taking care of the evolution ability and long-term commitment of a platform, as well as establishing a different mindset

- **Cloud-native platform**: Focusing on the abstraction and enablement to use standardized APIs and integrations

That perspective might feel fine-grained, but the term platform itself often leads to more confusion. A cloud platform is also a platform, right? A **Software as a Service (SaaS)** could also be seen as a platform. Referencing a cloud-native platform or IDP gives the right direction and understanding. Depending on your organization's maturity, it is therefore also essential to clarify these terms and establish a common understanding, language, and shared knowledge.

Platform case studies and success stories

To highlight the positive impact a platform can have, we can look at three totally different companies and their results from using IDPs. All of these cases primarily focus on Backstage as the developer portal and entry point for the IDP.

Spotify, as the inventor of Backstage and the mother of the IDP movement, claims that the following is true of their internal Backstage users:

- 2.3x more active on GitHub

- 2x more code changes

- 2x more deployments

- Onboarding time for new developers dropped from 60 days to 20 days

The Expedia Group reports different numbers:

- It takes four minutes, on average, to create a new component or app

- Over 4,000 users are using the IDP for at least 20 minutes per day

- The technical documentation is viewed over 50,000 times per month

- Just over 15% of the internal developer tools are integrated with Backstage, already reducing context-switching

Now, the last company we should take a look at is Toyota:

- Projects ship now weekly artifacts instead of monthly

- 8-12 weeks are saved on overhead efforts per team, resulting in over $5 million in reduced costs or time and budget used for value creation

- Standardize deployment templates reduce failure and speed up deployment

All those numbers are interesting to understand in the context of a digital-native company, a travel technology corporation, and one of the biggest car manufacturers. Any of them can show a clear positive effect *[2]*.

Projects versus products

Speaking of organization, introducing new solutions is commonly done as a project. So, at some point in time, someone decided to invest money in building their own platform. This approach faces one fundamental problem: the deadline. Projects are required to reach a target within a given time and budget frame. If the project runs out of time or money, it focuses on the operation and maintenance of it. These two parts of a life cycle are treated as separate things, causing a time period of rising sun and sunset. To explain this a bit, you can see heavy investments, communications, and excitement during the implementation phase. However, after hitting the deadline, the project turns into a dead object that requires maintenance. DevOps didn't change this behavior; it simply often got new names, new roles, and different processes. However, in the end, budget, people, and attention are turned away to the next project, while just a fraction of the former budget stays. This is frustrating for engineers who have worked hard on the implementation, and it will become frustrating for the organization over time when costs for pure maintenance keep increasing. Still, the people who have built it might leave or join other projects. This short-term view on implementing systems has slowly killed many good projects and team spirits. More importantly, it shows that the business value of the solution isn't clear. When an implementation, such as a platform, can provide explicit value, there shouldn't be a reason to turn away attention and cause its sundown.

While doing many implementations as projects is a valid approach, this is its death sentence with platform implementations. Regular implementations are feature complete; they can exist after they are done. But a platform will be always moving, always being upgraded and always be implementing new features. When working with open-source and cloud providers, you will learn early how fast tools and software are in their own development cycle. Features, fixes, and security patches are continuously published. This is a significant challenge for larger organizations as they are still used to far slower release cycles. The upside of keeping up the speed of this rapid development is that you, as an organization, can profit from new features and capabilities frequently. It is an innovation driver and enabler, allowing you to implement systems in other ways and solve problems you might not be aware of. Is a problem that isn't painful for you a real problem? Organizations tend not to consider such things as issues since they are used to only identifying painful processes and approaches.

Let us look at an example. In the current year (2024), the European Union released a law to improve companies' reporting on their carbon emissions. On a high level, this also includes IT resources. Also, within the last few years, multiple open source foundations and projects have been started to bring transparency to the energy consumption of software. A year ago, we would only have been able to report very rough, highly estimated numbers when it comes to energy consumption for a data center, for a server, and with some manual processing, for a piece of software. Today, we can obtain fine-grained information for any application running on bare metal, hypervisor, or containerized within Kubernetes as tooling has evolved to provide this data. Public CSPs provide more and more insights into their own energy consumption. What can we expect for the coming year? We can expect even better numbers, including the regional carbon mix of the energy and end-to-end visibility and transparency of such numbers. With platforms and platform engineering teams, such transparency will naturally come over time. It doesn't require a project turning IT upside down. Calling out the demand for it will result in platform engineering teams implementing those capabilities into the platform's core to benefit everyone who uses that platform to build, deploy, release, and operate software.

This is called a product mindset and it feels natural for platforms to adapt to the demands of their environment.

Platform as a product

Platforms as a product are user-centric, listening to and actively researching the end user demand to keep improving their services. A product is also aware of its value. Similar to any app on your mobile phone, it uses its own value to refinance further development and new features. Here, there are no deadlines and no sunsets. The goal is just to strive to become better with every release. What this gives to an organization is an expert team that keeps actively working on central enablement for providing your business with a platform of value generation.

Designing and developing a platform as a product goes beyond the pure engineering aspect. It faces organizational challenges that should be considered when you actively decide to build a platform. In fact, you have to deliver valid numbers on your benefits and show that your platform carries its own costs. This must be in the mindset of the product owner and platform engineers. The idea here is not to become business people but to be able to clearly communicate the reason for existence and, more importantly, to be a product that doesn't have a deadline.

Right now, you can find three different types of platforms as products:

- **IDPs**:
  - Provide a best-in-class experience for software engineers
  - Enable the development and operations teams for the end-to-end support and visibility of their software
  - Bring governance, compliance, and security
  - Establish a self-service for the development teams and simplify the deployment process
- **Data science and machine learning platforms**:
  - Similar to IDPs and often evolve out of those
  - Leverage their scalability to research, analyze, and process data cost-efficiently
  - Overcome complex implementation and make them generally available
  - Provide direct, secure access to relevant data sources

- **Low-code/business platforms**:

  - Strongly driven trend to provide platforms that bring solutions with which it is possible to implement new features with relatively less to almost no coding demand

  - We will see them more in the years to come

In our book, we will focus on the product-centric view of architecting IDPs.

Do you need a platform?

Like with any other complex environment, we first have to ask – do we really need a platform? Do we know what we will use it for?

Although platforms can provide a lot of benefits, they are not always the answer to your organizational questions. The signs that you are not ready for a platform yet are as follows:

- You have only monolithic applications

- You don't have your own development team

- Your DevOps, SysAdmin, or infrastructure team is heavily overworked or siloed

- You have very simple applications that can run anywhere

- You are having a hard time providing a budget for training to grow the skills of your teams

- You usually run commercial, off-the-shelf solutions

On the other hand, when does an IDP make sense to you? The following criteria are indicators that you're ready for an IDP:

- You have requirements for multiple infrastructure environments or foster a multi-cloud strategy

- You require advanced control over your environments (security, compliance, and deep insights into infrastructure and application behavior)

- Your development team is continuously overloaded with non-valuable tasks

- You have a curious and interested DevOps or infrastructure team that has taken its first steps toward a platform without knowing it

- Your application requires some kind of orchestration due to microservice architecture because many components or different integrations need to play well together

- You want to enable your organization to optimize your IT for costs, transparency, quality, or security

Before you define a platform product for your organization, you should answer all the points on the checklist. It makes sense to have multiple points illustrating why you need it. For example, having a team ask for an IDP, or having someone mention that they've heard of it in a conference, is not a strong foundation for making such a decision. The introduction of platforms is a journey, and from our experience, it can become the central focal point for one's company in a relatively short amount of time. Under such pressure, you still need to have a purpose and direction.

Now that we have learned how to define the purpose of our platform, we will need to discuss the question of whether we really need this additional abstraction layer.

Do we need yet another abstraction layer?

Let's briefly summarize what we have seen so far. A platform puts a bracket around, and a layer of abstraction on top of, your existing infrastructure and environments. The platform enhances it with further capabilities so that your development teams can utilize it in an automated, self-serving way.

From a technical perspective, this represents the next layer of abstraction. Therefore, it is only right to discuss this new fabric we put on top. Going from the bottom to the top, we see the bare metal, followed by the hypervisor for virtualization; this is topped by cloud providers. Some might include containers, Kubernetes, or serverless components – and now, we will add our platform. These are at least four layers, each promising to make the layer underneath simpler and glued together by a hard-to-be-defined meta-level of scripts, **Infrastructure as Code** (**IaC**), cloud libraries, and automation. So, do we really need this yet another layer, or are we using it to keep ourselves busy building things?

Declutter the abstraction layers

There is no simple answer to that question, but by looking into the purpose of each layer, you might be able to grow your own understanding of it.

Hypervisors were initially introduced to simplify the supply of hosts so that software could run and better utilize servers. Today, they still serve the same purpose but could be replaced by Kubernetes and container runtimes. A key argument against this replacement is that a virtual machine provides better isolation and higher security. Without getting too deep into this discussion, there are options to provide very solid isolation, such as with Kata containers. The only component that causes headaches is the container with the OS and its runtime. Looking some years into the future, **WebAssembly** (**Wasm**) could be one part of the answer to that problem. Without an operating system in the container and pure naked binary files, there are almost no gates open for attacks. However, let's give it some more time.

Infrastructure as a Service (IaaS) providers and public cloud providers enhance this with software-defined storage and networks, reducing the complexity of building your own data center and managing all physical dependencies. In addition, they provide further capabilities of commonly used scenarios, such as databases, load balancing, user management, message queues, or ML playgrounds and pre-trained AI models. This is a very useful implementation, which leads to a rapid development of the industry and an extension of what is possible. However, this also moved the whole industry in a problematic direction. The technology is developing faster than people and organizations, in particular, are able to adapt to it. We see that there is a shortage of professional engineers across the globe, while businesses are looking into providing more digital services every year. Solutions and their dependencies are therefore built natively to the cloud. The return on this effort can be significant. You are able to manage any kind of infrastructure and services with a relatively small group of people globally. Yet the reality also states that the average company has between two and three IaaS and public CPSs, plus (often) its own computing capacities, as well as around 10 SaaS providers, which can go up to around 50 for major enterprises *[1]*. CSPs also have between 40 to 200 different services. In other words, we are able to achieve a lot today, but the complexity of those environments has also become significant.

To tame this scale, IaC and **Cloud Development Kits (CDKs)** have become the tools of choice to manage your landing zones and software integrations. The fun part of the story is that practices such as DevOps, which are commonly misinterpreted, have made things even worse. These misinterpretations have now led to the sudden expectation for developers to also set up and maintain the infrastructure for their software needs.

Last but not least, we have systems based on containers, Kubernetes, or serverless. Dozens of options exist for each of them to provision those environments, deploy the code, and run the components. It's understandable that there are too many layers you have to take care of. However, their development is reasonable, as you don't want to do things in the old way of getting software up and running. Pushing code to images and from there to a runtime you choose simplifies the provisioning process.

Overall, to represent the level of complexity, we can think of a three-dimensional object such as a cube made of cubes. The following illustration shows the different service layers, representing different maturity of abstraction and how layer after layer comes together to form an IT environment.

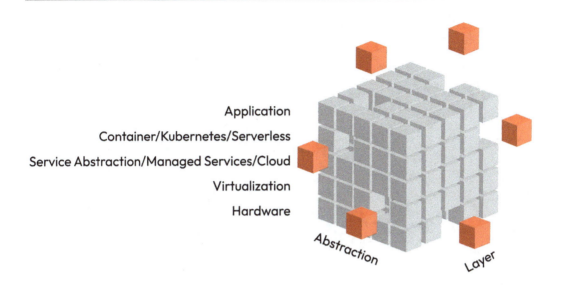

Application
Container/Kubernetes/Serverless
Service Abstraction/Managed Services/Cloud
Virtualization
Hardware

Abstraction

Layer

Figure 1.3: The multi-dimensional complexity of computing abstraction and simplification

Now, that figure is oversimplified, considering the hundreds and thousands of options you have in each dimension. However, it still gives a first good hint: if you need a platform, build a wrapper around this construction and tame its immense complexity to harness its power.

The cognitive load for software engineers and other IT professionals

In order to manage all these layers, we need to know about and use many tools, as well as follow various processes. It becomes difficult to focus on the actual job and create value while spending a large amount of time on things that should simplify our work. This is called the **cognitive load**. Originally made famous by Daniel Bryant, that term puts a bracket around the job overload and mental stress of many developers, as well as other specialists within IT. Reducing the cognitive load brings more happiness and satisfaction, but also effectiveness and reliability, to the engineers. Looking at the following graphic simplifies the perspective on what needs to be handled as a professional across the different decades. However, going forward, we have to reduce this load. AI could be part of this, alongside new concepts for running computing processes, and platforms of course.

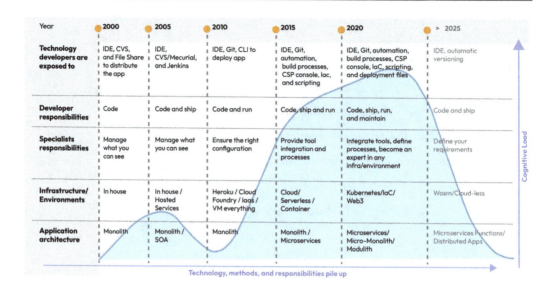

Year	2000	2005	2010	2015	2020	> 2025
Technology developers are exposed to	IDE, CVS, and File Share to distribute the app	IDE, CVS/Mecurial, and Jenkins	IDE, Git, CLI to deploy app	IDE, Git, automation, build processes, CSP console, IaC, and scripting	IDE, Git, automation, build processes, CSP console, IaC, scripting, and deployment files	IDE, automatic versioning
Developer responsibilities	Code	Code and ship	Code and run	Code, ship and run	Code, ship, run, and maintain	Code and ship
Specialists responsibilities	Manage what you can see	Manage what you can see	Ensure the right configuration	Provide tool integration and processes	Integrate tools, define processes, become an expert in any infra/environment	Define your requirements
Infrastructure/ Environments	In house	In house / Hosted Services	Heroku / Cloud Foundry / IaaS / VM everything	Cloud/ Serverless / Container	Kubernetes/IaC/ Web3	Wasm/Cloud-less
Application architecture	Monolith	Monolith / SOA	Monolith	Monolith / Microservices	Microservices/ Micro-Monolith/ Modulith	Microservices Functions/ Distributed Apps

Technology, methods, and responsibilities pile up

Cognitive Load

Figure 1.4: The extended cognitive load with a projection to an ideal future

Not only does technology change over time but it also piles up. This means that we have to run and maintain legacy systems while changing architecture styles and introducing new programming paradigms and new tools. This also changes the responsibilities and extends them far beyond the typical borders of one's job description from some years back. Breaking down this problem can reveal an answer to the question of whether we need a platform to solve our problems or whether implementing one will increase the complexity again.

In the end, every organization is different. Some are stuck in the early 2000s, and others continuously try to adapt to what comes next. Even within the same organization, you can often find drastic differences. One department might run everything on some VMs in their own data center, while the next might deploy functions within a global CDN or edge provider. Therefore, it's on you to draw up a vision, strategy, and goal for your platform, if you really need it.

The complexity we experience and the act of putting pressure on engineers needs to be encountered because IT tends to become more complicated over time. In the upcoming section, we will focus on implementing the right solution for developers to overcome that troublesome direction we are heading toward, and which can even lead to burnout.

Implementing developer- and product-focused solutions

Throughout the next few years, we will see an evolution of cloud computing. In this context, platforms will play a crucial role. On the one hand, the cloud will be everywhere, becoming an abstraction for infrastructure. It doesn't matter whether this is in the form of edge computing or very specialized services or offers. On the other hand, as we have learned in the previous section, we have to focus on delivering environments that enable the best experience possible for developers and other roles, so those people can focus on generating value. Bringing these elements together practically is the key enablement for an IT organization to keep up the speed with the market while delivering continuous value to your company.

The pervasive cloud

The **pervasive cloud** is not a single solution. It clusters a variety of cloud-computing capabilities that are undergoing a transformative shift to drive business and innovation significantly. The key advancements focus on the integration of cloud technologies anywhere, from private data centers over distributed computational networks to the edge. However, the pervasive cloud goes beyond that. It follows concepts to bridge physical gaps through sensors, IoT components, mobile devices, and other smart connected solutions. Therefore, it is known under other terms such as **ubiquitous computing**, **ambient intelligence**, or **everywhere**.

Gartner, the research company, assumes that six further technologies will shape the pervasive cloud and define its nature *[3]*:

- **Augmented FinOps**: Combines DevOps methods with cost optimization and budgeting
- **Cloud Development Environments (CDEs)**: Simplify and unify the development environment, reducing human errors and ensuring reproducibility
- **Cloud sustainability**: Achieving environmental, social, and economic benefits, reducing the harmful impact of the strongly growing cloud computing tech, and leveraging its power for good
- **Cloud-native**: Implementing cloud characteristics as defined before
- **Cloud-out to the edge**: CSP capabilities extended to the edge
- **Wasm:** The potential ubiquitous runtime and binary format for everywhere, but not necessarily everything

However, we need to ask why this is now relevant to us as platform engineers, architects, and developers.

First, you can find many of the technologies that we are already working on in these definitions and assumptions. Cloud-native, FinOps, edge, and CDEs are daily realities, while sustainable IT and Wasm have experienced heavy development in recent years. That's all relevant in making it clear that we are not discussing sci-fi technologies that won't be attainable within the next 100 years. It's happening right now and it is ready to be used. We develop and innovate all of those foundations; it just might not be as visible and prominent as GenAI.

Second, to extract the maximum value from cloud investments, businesses must adopt automated operational scaling, leverage cloud-native platform tools, and implement effective governance. These platforms integrate essential services such as SaaS, **Platform as a Service (PaaS)**, and IaaS to create comprehensive product offerings with modular capabilities. IT leaders are encouraged to utilize the modular nature of these platforms to maintain adaptability and agility in the face of rapid market disruptions. Imagine the complexity of such environments without a platform that tames this wide range of motion. Even so, with all that complexity, we need to keep the product mindset in focus, or else it will be hard to provide reliable IT services and solutions in the future.

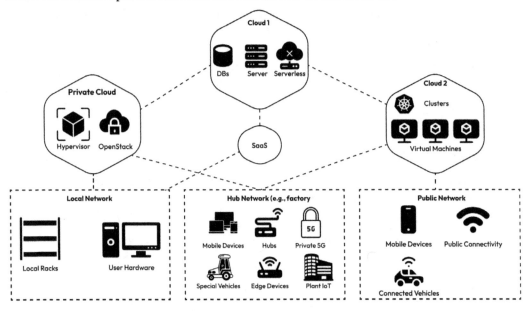

Figure 1.5: Cloud concepts are found everywhere in a pervasive cloud

Looking at the preceding diagram, you can find elements of the pervasive cloud everywhere. We shouldn't look at this figure as if those are separate items. Everything is connected. Apps on phones talk with services in the cloud or in local hubs, corporates have multiple networks connecting various computing environments with each other, and we have entirely skipped more progressive concepts such as **Web3** here.

> **Important note**
>
> IT as we know it today is undergoing a heavy transformation, both in the visible and invisible spectrum. With every step we take, we increase its complexity while facing demographic pressure and a shortage of professionals. Sooner or later, most companies will be required to have their own platform. If they don't, they will buy it as a service.

Focusing on developer experience

It is not sustainable to hope that every developer will be able to cover the extremely wide landscape of tools and technologies without burning out within a few years. Therefore, the quality of user experience is pivotal in determining the adoption and success of a platform. A well-designed platform means that it is intuitive, easy to navigate, and aligned with the developers' expectations and workflows. Enhancing the experience involves streamlining interactions, minimizing friction points, and providing a visually, technically, and functionally pleasing environment. This not only improves user satisfaction but also boosts productivity and engagement. The question is how to achieve this.

We must consider that every developer might have a different preference when designing the platform. It starts directly with the problem of the interaction between the platform and the user. Developers might ask various questions, such as the following. Do we need to set up a portal? Is pushing code on a Git service enough? Can I interact with the platform via CLI? It can be hard to tell, but successful platforms provide all of those interactions. Starting with an API-centric approach will enable any other path to be taken simultaneously. A strong API is the core of a good platform. In reality, most platforms still provide multiple different interfaces. The rapid development of tools to unify this will overcome such challenges and if considered to be built on greenfield, it can be then placed directly into the core.

An example of such a core is Kratix. The Apache 2.0-licensed open source platform describes itself as a "... *platform framework for building composable IDPs.*" In the following figure, you can see how Kratix positions itself between all the common tools we use today and provides one entry point.

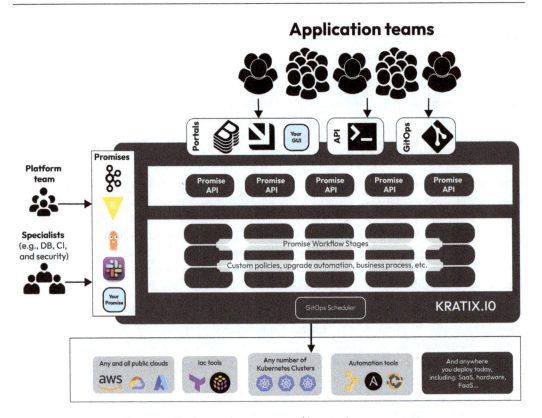

Figure 1.6: Kratix overview as a central integration component

Kratix achieves this through the concept of *Promises*, which is technically a YAML document that defines a contract between the platform and the users. Every team has to go through a complex onboarding process, not because of the platform itself but because of other dependencies such as CI/CD, Git repositories, and linking everything together. With Kratix Promises, you encapsulate all those steps or combine multiple Promises into one.

Now, Kratix supports simplifying the platform foundation for the developer experience, yet something is missing. The other side of the coin is a developer portal. Backstage is an example of an open source Apache 2.0-licensed solution developed by Spotify. Kratix and Backstage are working well together and integrating seamlessly. Backstage is a framework that enables GUIs to be declaratively created with the aim of unifying infrastructure tooling, services, and documentation to produce a fantastic developer experience. Backstage comes with three core features: the service definition, the Backstage service catalog, and its plugin system, through which you can enable further features such as docs.

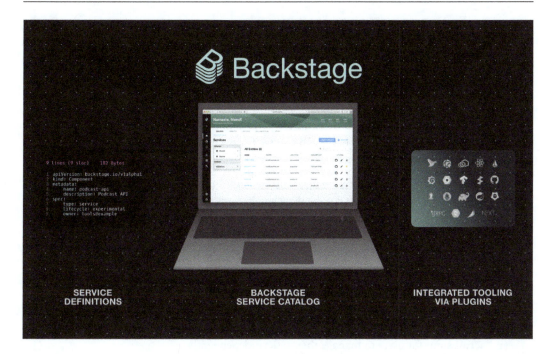

Figure 1.7: Backstage's three core features

At this point, we have seen the challenges that need to be solved, and we have taken a sneak peek into the solution space. That should give us a feeling of the current possibilities before we dive into details throughout the next chapters.

Attributes of platforms

A platform must fulfill certain attributes and provide some core components that allow us to serve capabilities to the end user. So far, we have learned about all the complexity to handle, the integrations to take on, and the focus and mindset on the end user to create the best possible experience and product. All of this must be matched with a technical, processual, or methodical approach in the shape of attributes. For the development of an idea for your platform, we can only encourage you to look into attributes and then decide on the best solution for you, rather than taking a solution and constructing its usefulness around your case.

Some attributes, such as reducing the cognitive load, the platform as a product, or the developer/user experience, are well covered. However, there is more to be considered:

- **Flexibility, adjustability, and composability**: Platforms should offer flexibility in how they are used and integrated with other systems. By supporting modular and composable design, users can customize and extend the platform with optional features that suit their specific needs without being overwhelmed by unnecessary functionalities. This approach allows the platform to serve a broad user base with diverse requirements while maintaining simplicity and manageability.

- **Secure**: Security is a critical attribute that must be embedded in the design and architecture of the platform. **Secure by default** means that the platform employs the best security practices and configurations out of the box. Users should have robust security measures in place without needing to configure them extensively. This includes data encryption, secure access controls, and regular security updates to protect against vulnerabilities.

- **Self-service**: Enabling a self-service functionality allows users to perform tasks such as setting up environments, deploying applications, and accessing services without waiting for IT support. This capability not only accelerates workflows but also reduces the operational burden on platform teams. Self-service portals should be user-friendly and provide all necessary tools and permissions for users to manage their tasks independently.

- **Documentation and support**: Effective documentation and onboarding are essential for empowering users and reducing the initial learning curve associated with a platform. Comprehensive, clear, and accessible documentation ensures that users can self-serve to solve problems and understand the platform's capabilities without external help. Onboarding processes should guide new users through the platform's core features and functionalities, making them feel competent and confident in using the platform from the start.

As you might have realized, those attributes, in some way, match the outline of this book and are therefore our lighthouses on the way to designing and planning a platform.

Sometimes, such attributes might be too soft to be used as a reference to start an implementation. An alternative for this was described by the internal developer platform community (`https://internaldeveloperplatform.org/`) as the five core components of an IDP [4], which we encourage you to check out as an alternative source:

- **Application configuration management**: Manage the application configuration in a dynamic, scalable, and reliable way

- **Infrastructure orchestration**: Orchestrate infrastructure in a dynamic and intelligent way depending on the context

- **Environment management**: Enable developers to create new and fully provisioned environments whenever needed

- **Deployment management**: Implement a delivery pipeline for continuous delivery or even continuous deployment
- **Role-based access control**: Manage who can do what in a scalable way

In *Chapter 2*, we will go into detail about the organizational and technical aspects, as well as the development, of an implementable plan for crafting your platform as a product.

Focusing on the developer perspective means to impact the organization and with that, the people who are working for it. Next, we will look into the sociotechnical aspect that we have to consider while creating platforms.

Understanding the socio-technical aspects

Designing a successful platform involves much more than technical prowess; it requires a deep understanding of its users' diverse needs, fostering and motivating collaboration from an early stage, and creating an open environment that welcomes a platform-centric culture. These sociotechnical aspects of platform engineering are crucial perspectives that emphasize not just the technical components of a platform but also the human elements—how individuals and groups interact with the system and how it influences their work and behaviors. Understanding this often invisible part of creating a platform is essential. It is the glue that defines that technologically robust systems are relevant, as well as defining the deeply integrated daily workflows and behaviors of the users. Respecting this almost meta level means enhancing productivity and driving adoption through high satisfaction. We have to acknowledge that every technical decision has consequences, both technologically and for humans. Therefore, it's important to design platforms that resonate well with their users, foster a collaborative culture, drive innovation, and just make work easier on a daily basis. By focusing on these sociotechnical aspects, platform engineers can create more adaptable, sustainable, and user-centric systems that stand the test of time in an ever-evolving cloud-native landscape.

However, we must be aware that we interact within a socio-technical system. The challenge is that we have continuous friction and optimization within our work, personal, and private environments. The work environment represents what's happening in your professional context. The personal environment is an "I"-centric perspective that shares the experience with anyone who is in touch with you, which can be work or private elements. Lastly, the private environment is what others can estimate and observe, meaning what happens behind closed doors to you but can have an effect on your opinions and perspectives. This constant moving and adjusting happens between the company or project structure, the people, the technology, and their tasks. Within this, we have different influences and motivational drivers. The following representation should visualize this continuous movement. Subsystems and systems themselves have their own rules, activities, and powers.

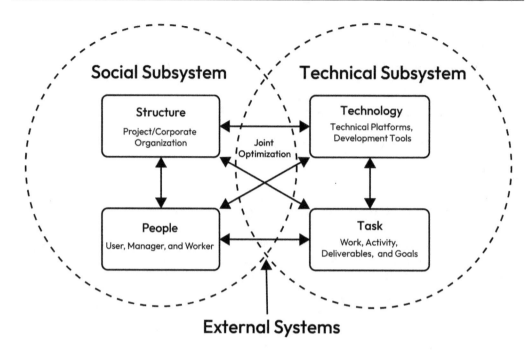

Figure 1.8: Sociotechnical system, Trancossi et al.

When we are working on identifying the best approach to implement a platform, trying to find out the needs and demands, building a community, and advocating for openness, we must be aware that this all is influenced by and happening within a socio-technical system. We might sometimes be blocked or pushed into a different direction, or find it hard to motivate people. That's the right time to spend some effort on understanding the subsystem and external systems that are pushing you.

Understand user needs in platform design

A platform design must inherently be flexible to accommodate a wider range of stakeholders than only developers. We can't say it enough, but implementing a platform is an all-hands-on-deck decision. Software engineers, product owners, business stakeholders, and operative teams all have unique requirements and challenges they expect the platform to handle. Developers may seek straightforward tools for deployment and testing, while end users need intuitive interfaces and seamless interaction. Business stakeholders, meanwhile, are likely to be primarily focused on **Return on Investment** (**ROI**), security, governance, compliance, and scalability. As a platform engineer and architect, it is your responsibility to identify these demands and balance them throughout the platform's lifecycle.

So, the first step in user-oriented platform design is to accurately identify and understand the different stakeholder groups interacting with the platform. In the early stages, you can do interviews and surveys to get a clear picture. Consider a 360-degree view and know that it's better to interview one person too many than to interview too few people overall. In later stages, it's helpful to be more data-driven and to use usage data for the different components of the platform. Service requests and open problems and issues at the help desk will be a good indicator of the useability of the platform. At this point, it becomes relevant to see the platform as a product and to have a product mindset, as you have to consider user feedback from the earliest stages of development and throughout the product life cycle, ensuring the platform is accessible, intuitive, and efficient. A false sense of vanity will definitely steer your efforts in a bad direction. It is helpful to define principles such as transparency or simplicity that will guide you through the development and decision process. In the upcoming chapter, we will tackle this topic in depth.

To remain relevant and efficient, platforms must evolve. Remember that projects die when they hit the deadline; products evolve with every release and piece of user feedback. Establishing robust release documentation and feedback loops is critical for this iterative process of improvement. Which channel suits your organization and users best is for you to decide, but there are many ways beyond surveys and interviews. Examples include direct interactions via communication tools, user forums, support solutions, even internal corporate social media, or in an extreme case, embedded feedback mechanisms within the platform itself. Regular updates and upgrades, defined by real user experiences and challenges, ensure that the platform remains aligned with user needs and industry standards, fostering loyalty and sustained engagement.

Foster and enhance collaboration

The issue of transparent, open feedback loops is that they require a well-set-up collaboration. Otherwise, it feels artificial, and in some sense like the traditional approach of requirements engineering. As platform engineers, we have to package these classic methods into a more personal and welcoming approach. Effective collaboration is the cornerstone of any successful platform. By integrating the right communication tools, creating a fantastic user experience through customization, and advocating for cross-functional teamwork, platforms can become more than just technology solutions—they can become a place for collaboration between different development teams and a driver for organizations and innovation.

To foster collaboration, your team and the platform must be open and welcoming. Reduce any kind of barriers to onboarding a new product team and provide them with differently presented starting points. It must be clear how and where to reach your team for any kinds of questions. This is important in the early days of a new end user team joining your platform. Eliminate any kind of implicit expectation that it is clear where and how to find your team, as it usually isn't, especially in larger organizations. This requires that you and your team become platform advocates, alongside internal public documentation and landing pages. You have to go out there, work with other teams together, listen to them talk about their challenges, and show them your solutions. Where possible, you have to inform your organization about new features and use cases, and keep showing them how simple it is to get started.

Cultivating an open, platform-centric culture

Cultivating an open and platform-centric culture might be the hardest part. It requires buy-in from other budget-managing roles within your organization, in addition to your engagement. Cultivation is based on training, motivation (which may be incentivized), and engagement with your community or user group. All of these activities go beyond your budget responsibility.

Training might look like an expensive, time-consuming activity that quickly becomes outdated, but there is almost no better way to get into close and truly open contact with your end users. Comprehensive training programs combine different approaches and sources:

- In-person workshops that focus on different components of your platform
- Hands-on sessions for onboarding and a good first project or specific integration that is less complicated
- Do-it-yourself tutorials
- Online/video tutorials for generics
- Online/video tutorials for your specifics

As you can see, you don't have to do it all yourself. Buy courses and training for common knowledge and provide your own to deepen this knowledge and teach your platform specifics. Additionally, consider creating a repository and wiki of resources such as FAQs, best practice guides, troubleshooting tips, code snippets and examples, use case recommendations, and post-mortems. This can help users feel more empowered and reduce the learning curve for new technology adoption.

To keep up the motivation, the training and platform usage can be combined with incentives. There are different motivational strategies such as gamification, recognition programs, and performance-based rewards that can encourage active use of the platform. For large organizations, we have even seen internal badge programs and *cloud driver licenses*, which could always be combined with official certifications. Yet at the same time, you need to be careful not to create artificially high barriers. That would be counterproductive. Ideally, your incentives should motivate new users and provide challenges for experts. For example, they can be integrated with existing performance metrics to reinforce a culture that values continuous improvement and effective use of technology. In this way, you can also influence corporate targets through defined KPIs, such as cost reduction, performance improvement, and stability.

Finally, we have to combine all those approaches. The best way to do this is to create a community. We can establish such a community through user groups, regular meetups, and forums that enable users to share knowledge, solve problems collaboratively, and provide support to each other. To nurture a community, we highly recommend that you find support from your C-level and marketing team. It might sound easy to just bring people together, and as though all will be well once that happens. In truth, it is a whole discipline that requires exploring strategies for engaging community members, facilitating productive discussions, and fostering a sense of belonging and ownership among users.

Without people and respecting their needs, your platform will either quickly be abandoned or never gain any traction in the first place. Strategic initiatives focused on training, incentives, and a thriving community can accelerate an open, platform-centric culture. To increase a platform's effectiveness, you have to go beyond its pure technical deployment. This holistic approach ensures that the platform becomes a fundamental part of the organization, driving innovation and efficiency across all levels and stakeholders, from developers and operations to business and process owners.

Summary

In this opening chapter, we have explored the role of platforms in modern software development, particularly within cloud-native environments. We've established that platforms are not just infrastructure elements but elemental products that require strategic planning and continuous refinement to align with evolving technological and business demands.

You have learned about the platform fundamentals, and you recognize the comprehensive nature of platforms that combine software development, operation, and deployment into a cohesive environment. To build on that, you saw why the platform as a product mindset is a key component. It entails continuous development, user engagement, and responsiveness to feedback, to maintain relevance and value.

Then, we discussed the importance of simplifying the developer's interactions with the platform, significantly reducing cognitive load and operational complexity. Two short tool examples should cause your curiosity and interest in the following chapters but also show you that all those complexities and challenges are solvable.

In the final part, we learned about the socio-technical aspects and the obvious relevance of the human factor. Platforms must consider both technical capabilities and human factors, ensuring that they support the workflows, collaboration, and productivity of their users. For this, we introduced the three pillars to build on: understanding the user needs before designing a solution, fostering and enhancing collaboration, and cultivating an open, platform-centric culture.

These lessons should motivate you to dive deeper into platform architectures and how to build them, as they enhance operational efficiency and foster innovation and agility within organizations. By internalizing these concepts, platform engineers and architects can design solutions that are robust, user-centric, and adaptive to change, while delivering true value for their organizations.

Moving on from the people aspects to the architecture, we will discover how to define the principles and purpose of a platform, how to define the platform architecture, and how to measure the platform's success in the upcoming chapter.

Further reading

- [1] CNCF Annual Survey 2023: `https://www.cncf.io/reports/cncf-annual-survey-2023`

- [2] Case studies on IDP success:

    - Spotify: `https://engineering.atspotify.com/2024/04/supercharged-developer-portals/`

    - Expedia: `https://backstage.io/blog/2023/08/17/expedia-proof-of-value-metrics-2/`

    - Toyota: `https://backstage.spotify.com/discover/blog/adopter-spotlight-toyota/`

- [3] Gartner 2023 hype cycle for emerging technologies: `https://www.gartner.com/en/articles/what-s-new-in-the-2023-gartner-hype-cycle-for-emerging-technologies`

- [4] *The 5 Core Components of an Internal Developer Platform (IDP)*: `https://internaldeveloperplatform.org/core-components/`

Get This Book's PDF Version and Exclusive Extras

UNLOCK NOW

Scan the QR code (or go to `packtpub.com/unlock`). Search for this book by name, confirm the edition, and then follow the steps on the page.

Note: Keep your invoice handy. Purchases made directly from Packt don't require one.

2

Understanding Platform Architecture to Build Platform as a Product

In this chapter, you will be guided through all the relevant groundwork and approaches to create your platform architecture. First, you will learn about the principles and how they become a guiding part of your platform strategy. This will help you to stay on track and focus on what matters most. We will round this up by defining the purpose of your platform.

From here, we will begin the process of defining your architecture. You will learn about the components of a platform, platform composability, and how to handle dependencies. Next, you will work on reference architectures and how to create them for your demands. To deliver new perspectives to you, we have gathered a couple of platform-as-a-product use cases.

Last, you will learn how to start your platform journey with the **Thinnest Viable Platform** (**TVP**) and how you can make platform adoption visible through relevant **key performance indicators** (**KPIs**).

What you will learn in *Chapter 2* is as follows:

- Understanding platform principles and defining the purpose of your platform and team
- Exploring platform architecture – layers, components, and meta-dependencies
- Exploring platform as a product – use cases and implementations
- Understanding TVPs
- Looking at the relevant KPIs to make adoption transparent

Understanding platform principles and defining the purpose of your platform and team

You need two primary things to develop the right architecture for your platform and follow it. First, you have to define the guardrails that will guide you in your direction and the steps taken to develop and run your **Internal Development Platform (IDP)** as a product. Second, you need to understand what you are aiming for. This is not meant as architecture, but as the identified and defined purposes that are standing behind your platform.

Introducing principles as guardrails for decision-making

Principles might sound like an old-school approach from the forgotten times of enterprise architects and very stiff organizations. In times when everything becomes agile, principles are the cornerstones for active decision-making. They help cut discussions short and focus on the platform delivery. How are principles different from capabilities or attributes? A platform's capabilities are the functionalities an IDP can provide or offer. Think about it like you require file storage, and with the specification of a resource definition that you apply within your platform, the system provides you with an S3 compliant bucket. Attributes are a superset of provided capabilities and are actively or passively consumed. If you define a secure platform, that is the attribute, then this might combine the hardening of the nodes, **Common Vulnerabilities and Exposures (CVE)** scans, and things such as network encryption and strict **role-based access control (RBAC)** rules. A principle helps you decide HOW you are going to achieve those capabilities and attributes. Therefore, you will usually encounter two types of principles:

- **General principles**: These are organization-wide counting orientation points. A common one from recent years is this: *Data is an asset*. This translates to the following meaning:

  All data is valuable to an organization. It's a measurable resource to which we apply practices to protect and use it:

  - For each data source, someone is responsible

  - Data is made available to everyone

  - We will protect data from loss and security threats

- **Specific principles**: These are use-case-oriented principles that can be relevant to others but are meant to target just a small group of people/specialists. An example from the security branch is this: *Any activity is defined by an identity*. This can be interpreted as follows:

  - For every process in a system, an identity is logged

  - Unknown identities are immediately to be reported

  - Every system requires user authentication

Having those two principles in mind will drive your decisions in different directions when you design the architecture for your scenario. What you will encounter is that specific principles look like they describe the obvious expectations of a department. But never assume or implicitly expect something if you want to be sure that it is done the way you want it.

> **Important note**
>
> Define *explicit* rules, guidelines, and frameworks if you have certain expectations of the outcome of one's activities. *Implicitly* requires that one can read your mind and have the same worldview as you.

Examples of platform and platform engineering principles

Before moving over to helping you define your own principles, we would like to give you some input on possible principles:

- **Focus on common problems**: Identify and prioritize problems that when solved help many departments or people in your organization. Prevent others from reinventing the wheel by eliminating shared issues and snowballing problems that, when cascading down, become bigger and bigger.

- **Don't reinvent the wheel**: Research the market for potential solutions that solve your demands. Periodically recheck homegrown solutions for standard replacements. The market develops faster than we do. Focus on creating value, not internal dependencies.

- **Each customer is important**: Every user of your platform is relevant, from large systems to small products. Gather and evaluate every feedback equally. Address everyone's needs and demands in the same way and provide the same class of service. Consider those inputs as the primary way to define your development path.

- **Eliminate waste**: We will build our platform to be as resource-efficient as possible and help our customers be as optimized as possible. We drive transparency on resource consumption and provide options to scale, reduce, and rightsize workload. Together, we develop best practices for developers to understand how they can configure and use advanced technologies for more flexibility.

- **A platform is a product**: Recall your product mindset for every decision. Focus on what provides real value to your internal customers, such as developers, operations, security, and others. Ensure features ship fast and have a short feedback loop. Recognize distractions from non-value-driven technologies. Your highest goal is to make your customers want to use your platform, not to force them.

- **Inner source**: Every line of code of your platform is accessible to your internal organization. Issues and problems are tracked publicly and their resolution is documented. Where relevant, maintain a FAQ. Everyone can propose a feature, a bugfix, lines of code, and documentation to improve your platform. Drive community events and foster exchange between you and all users.

Those principles try to cover some different viewpoints and levels of detail. Which is the right one for you is difficult to say. Don't be afraid to adjust wording or the principles over time if you recognize that they don't work.

> **Product mindset as the core principle**
> Your internal customers should want to use your platform. They shouldn't be forced to.

Defining your own principles

After you adapt to the idea of principles, all of them look good, right? Now, the challenge is to define and write them in a way that will be adapted by your team. They need to fit into your wording and your organization's internal approach to defining similar elements.

Principles can be derived from your corporate values, customer feedback, or a team's internal workshop. Whatever you do, it is important to have transparency about the origin of the principles; otherwise, the team may feel excluded. You will also need to iterate over them a couple of times to sharpen the wording and boil it down to the elemental things. As with all principles, they shouldn't be too long, but they should also not be too short. The sentences should be structured short and crisp without any nesting or conditions. Conditions are either a sign that you are unsure about the principle or that it can be two principles. The right length is when you address one topic, give it two or three explaining sentences, and maybe a motivation or rationalization.

I have seen two ways of writing that are very efficient in defining a principle and finding frequent quotations and adaptation, and sometimes, it's in Slack/Teams channel descriptions or email footers.

Command and rationalize

The **command and rationalization** approach usually works well for organizations that have stronger hierarchies, where the platform team may have just been formed or where the whole initiative is looked at with a skeptical eye. This requires a more strict direction for initialization. However, I believe that those principles should be reworked over time and the input of the platform engineering team and end user should be gathered.

Here is an outline showing how to write it:

- **Title**: Every principle has something like a title that should stick in one's head.

    The title should be followed by the list of commands:

    I.   The initial command sets the direction

    II. Then, there should be a direction on why that's relevant

    III. After that, there should be some more detailed actions to consider

    IV. Then, there should be another action or rationale

    V.   Here, the responsibility should be handed over to the reader

Here's an example based on the preceding outline:

- **Title**: *Self-Service First*:

    I.   Every platform capability must be utilizable by the end user without interacting with the platform engineering team

    II.   This enables the self-determination of the user and offloads work from the platform and DevOps teams

    III.   Don't prescribe specific ways, but guide through best practices

    IV.   Open the possibility of full transparency of your end user's deployments and allow fine-grained access to operational data

    V.   You are providing a highly valuable set of tools and automation that simplify the software's development and operation

As you can see, this style tells the reader what to do. The title is strong and says, *Whatever you do, first think about making it self-service available*. Take care with such number-style naming first. This doesn't work if you name every principle with first, primary, or overarching. However, here, you can see that from all of the potential principles you are writing down, this one will always be first, and will most likely overrule other principles.

Sentences II-IV detail and rationalize the first sentence. They provide supportive arguments and guidelines to consider. The closing statement is as important as the title or first sentence. This must be a handover statement. It is like a call to action, but not in a way that asks someone to do something – this is what happened in the sentences before. It must be addressed on a personal level, with a direct speech toward the reader. Have a look at the list of examples for other ideas on how to close a principle.

Platform teams will grow their internal trust and motivation over time, which demands a different style of principles.

We will …

Mature platform engineering teams and young organizations require principles that call their *WE* feeling. This is almost like a vow of brotherhood and sisterhood. However, you also need to know your team well and handle it with care. Not everyone likes this way of defining principles, as they are calling for strong bonds and communication. This is just a personal question and nothing negative, but you might lose valuable team members if you ignore it. This is how I would define a *We will* principle:

- **Outline**:

    Title: No changes to the title; bold, short, and crisp, that's it:

    I.   The first sentence usually combines the will of the principle and includes a help statement to rationalize it

    II.   The next sentences are all defining some more targets or approaches

    III.   Each sentence will start with "we," "together," "as a team," "for our customer," and so on

- **Example**:

    Title: *Self-Service First*:

    I.   We will provide our customers with every platform capability as a self-service, focusing on their **user experience (UX)** and enabling self-determined software development

    II. We provide best practices and cooperate in the definition of approaches that work best

    III. Together, we create a transparent, value-generating environment that fosters automation and end-user-centric capabilities to simplify software development and operation

What are your thoughts on this approach? I kept the principle the same so you can make a side-to-side comparison. Which one are you in favor of?

You can see that the *we will* is shorter, and a bit more complicated. That's because you want to address the team, tell them what to focus on, and provide a reasoning behind it in one sentence. Cutting those into short sentences wouldn't give the same feeling and might send the wrong message.

A combination of both approaches can work well and help you to identify which type suits your team better. However, it is not unusual for platform teams to be happy with the mix. It might feel heavy if you have four to six principles that all are calling your *we* feeling, but it may also feel very pressuring to have a list of commands.

Here is a summary of this in the following list:

- Principles are helpful guardrails in your architectural and engineering decisions
- Between four to six principles work best
- If you are unsure, start with less and involve the end user and the platform engineers to identify the next principles
- Frequently check your principles to see if they still suit your team, but don't change them every few months

Developing the purpose of your platform as a product

For a successful start to creating a platform, we need to define its purpose and target and how we can structure or restructure the platform engineering team. This step also helps you do a first iteration of your previously defined principles. Reflect on your own perception and challenge your initial definition. This will allow you to learn how to integrate two different forces and contradictory targets.

It might seem simple to start with a platform as a product. Changing the mindset, talking with customers, and providing some good features are all that you need. This is, however, only one part of the story. We have to make you aware that a cloud-native platform has the potential to become your organization's powerhouse of software development, delivery, and operation. Therefore, we will introduce you to different concepts that you can also find in other organizational theories and practices, such as Kanban, Lean Inception, or Team Topologies.

Understanding your own value

If humans are really bad at one thing, then it is in doing estimations. Without a scale or any reference points, we have a hard time estimating correctly. Say, I'm going to ask you, what is the value of a platform for your organization? You might come up with arguments such as a faster software development cycle, less operational overhead, and happy engineers that reduce recruiting costs. None of it is wrong, yet nothing about it is precise. In business terms, the value of a platform as a product lies in its enablement of time to market, or optimizing flexibility in the cost of delay. In other words: the longer you need to put a service out into the market, the less value you create. A good platform will allow you to optimize this value by providing high flexibility and speed to the implementing teams.

In the theory of **cost of delay** [1], it is assumed that the capacities are fixed. If you want to implement three projects, you need to decide with which you will start. Each project brings a different value. If you start with all three simultaneously, the assumption goes so far that you will finish with everything at the same time, but you are not faster or slower than if you would do them one after the other. With an IDP that fosters self-service, a high level of automation and autonomy, reliability, and so on; you reduce the dependency on the available capacities, allowing multiple projects to be implemented simultaneously. Thinking further, that's one of the reasons why public cloud providers are such a success. However, these benefits become an obstacle with rising complexity and wrong role definitions.

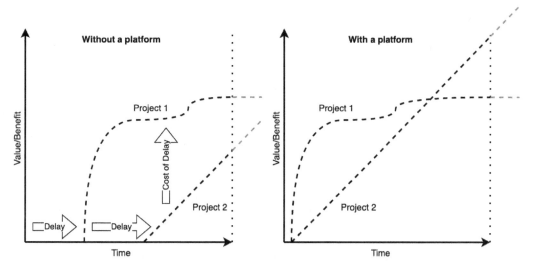

Figure 2.1: Cost of delay without and with a platform

The preceding figure shows the difference between implementing projects without and with a platform. On the left side, we can see at least two delays, which means the second project starts providing value late in the year or period. This delay can be caused by manual processes, dependencies on other teams and team members, or various technical steps that aren't integrated with each other. On the right side, a simultaneous start of the projects will result in a substantial value increase due to a very good platform. Remember that the area below the lines represents the value.

> **Important note**
>
> Cloud-native platforms, such as IDPs, are breaking the theory of cost of delay, allowing a fast and early value generation. The better the platform, the fewer dependencies on capacities you have, and the more efficient teams can implement and operate their products, the more value your organization can generate.

From a system to a platform as a product

In the end, a product is nothing more than an object with services made of your customer's demand, your *business* interests, some technology, and a holistic UX. If one of these elements is lacking, your product will fail. In a project world, we have three cornerstones constantly pulling on each other: time, money, and scope. This is known as the **iron triangle**. Finding a balance between those directions requires careful adjustments. But as we want to get rid of the project perspective, we will introduce you here to the product iron triangle. This consists of feasibility, desirability, and viability.

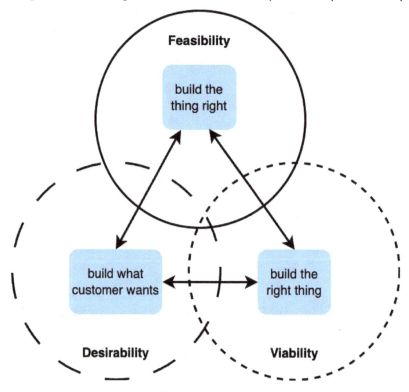

Figure 2.2: The iron triangle of products

We can find within that balance different forces pulling on the product. So, we have to ask ourselves the following questions:

- What need or job do we solve for the customer and user?
- What does our customer want? (Usually, they don't know.)
- How can we make it work?
- Which technology should we use?
- How can we make a sustainable business from it?

We can now combine this product perspective with a platform perspective.

"A digital platform is a foundation of self-service APIs, tools, services, knowledge and support which are arranged as a compelling internal product" Even *Bottcher* once described a platform like that *[2]*. Furthermore, he said, *"Autonomous delivery teams can make use of the platform to deliver product features at a higher pace, with reduced coordination."*

Platforms and products have a huge overlap. Both do the following:

- They must advocate for their usefulness and market it to internal teams
- They should be carefully designed and accurate
- They are designed with the user in mind or, even better, co-designed with users/devs (with a focus on UX/DevEx).
- They evolve offered capabilities with time and need a clear roadmap.

When developing a clear vision and purpose for your platform as a product, these aspects should be considered and checked. You will naturally find that for the product iron triangle, two roles will be immediately assigned.

The platform's product owner will handle viability, while the platform engineer will handle feasibility. But almost every organization lacks the role of the person who handles desirability. This role is commonly known as **Developer Experience (DevX)**.

Without a DevX, a platform engineering team is incomplete.

> **Important note**
> The platform's product owner handles viability. The platform engineers handle feasibility. The DevX handles the desirability. These roles define the success of a platform as a product.

Stakeholders and their impact on your backlog

Talking about stakeholders and customers in an inner corporate world unfolds a landscape of individuals with many interests. To focus on what matters, I recommend you categorize them into three groups:

- **Users**: The group of people who are really going to use your platform/product. In our case, it will be developers or DevOps.

- **Customers**: People who have budget power and/or make decisions for a whole team, department, or similar.

- **Influencers**: Other people who have an interest in your platform, such as security or operations teams.

You are building your platform for those stakeholder groups, not for yourself! While defining the purpose of your platform, you should look at the different stakeholder groups from different perspectives. Their needs might change over time, which you can find out with frequent iterations and probes. When you define features for the backlog, make clear from which stakeholders those requirements are coming. If some time has passed, usually six months is a good time frame, then you should reprioritize the feature requests with your stakeholders. This, on the other hand, will also impact the purpose of your platform. Let's look at an example.

Some while ago, we were asked to stabilize the customer's platform. We knew a little about the workload, but the overall system was unreliable and barely served as a useful system. Fast forward, after a couple of weeks of work had passed, new stakeholders had popped up. They became interested in the platform as it automatically adjusted its size to fit the workload, and people are generally happy with the fast and uncomplicated onboarding. Besides, a single tenant deployment for strict isolation was given during the former implementation. The new stakeholders were from a data science department and were looking for an easy approach to run their models without becoming experts. The cloud provider they used was already too complex, as they just focused on machine learning. The platform owner agreed to provide a couple of weeks of capacity to implement the tooling on top of the platform needed by the data scientists. Now, being far away from perfection, the implementation and platform were good enough to be easy to use, cheap, and ensured any data protection regulations. Fast forward again to the modern day, and the platform team's main purpose is to provide for data science teams' isolated environments, which come with a whole bunch of tools and are torn down when their job is done. Also, the platform can still run any other workloads, there is currently no demand for it.

This real-world example shows that sometimes the external influence is stronger than your own drive, but it can lead to higher acceptance and adoption throughout the organization.

> **Important note**
>
> In the end, you built a platform as a product for your stakeholders, primarily the users, and not for yourself.

Challenging Conway's law

```
Organizations which design systems are constrained to produce
designs which are copies of the communication structures of these
organizations. (Melvin Conway, 1967.)
```

I believe one of the top purposes of a platform is to do things better than they currently are. For this, we have to challenge an organization's status quo. Companies reorganize themselves continuously, exchanging responsibilities and adjusting their structures. But in the end, the organization will stay the same. That's Conways' fault – at least he described his observations of organizations. Platforms are here to disrupt and harmonize; utilizing, on the one hand, the DevOps methodologies to identify bottlenecks and disrupt those structures, and, on the other hand, harmonizing the efforts of the specialist teams into a unified environment.

As a platform team, you should first define your target and an ideology you want to follow to fulfill your purpose before you are forced to follow your organization's structure. Sadly, this leads to political games – how to gain more power, how to have the most users, and who has the best connections for driving decisions. But don't be too naive to believe that having a nice platform with some fantastic features will be enough to be successful.

Therefore, you will need a good definition of your team's purpose to keep the direction in stormy times, move on when everything wants to hold you back, or keep you in line when you have too much speed and traction. It's there to help you keep a balance.

To bring this all together and save it for your team and afterworld, you can use *The Platform Engineering Purpose Canvas* as a starting point. You can use the following template (`https://miro.com/miroverse/platform-engineering-purpose-canvas-template/`) or go ahead and create it on your own.

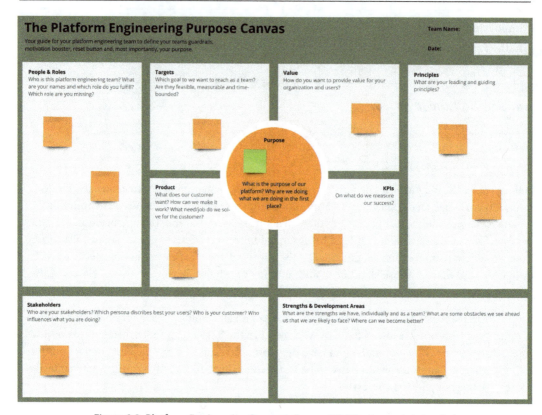

Figure 2.3: Platform Engineering Purpose Canvas [3] (The image is intended
as a visual reference; the textual information isn't essential.)

As we focus on designing and creating a platform as a product, we will not go any further into the details of platform engineering teams, how to set them up, and provide the right change management approaches. There are other books and methods that focus on this; therefore, we recommend that you check out *Team Topologies* if you are also interested in team development *[4]*.

Equipped with your platform purpose, we can take on the next challenge of creating your platform architecture step by step. In the next section, we will create a reference architecture and other diagrams for your platform as a product.

Exploring platform architecture – layers, components, and meta-dependencies

The navigation for your platform journey is defined by the reference architecture on which you can align your whole implementation. Architecture, and especially architectural diagrams, are not made for eternity. They are a framework, a blueprint you have to orient on, but they can and should be changed over time due to the evolution of the given market solutions, whether they are open or closed source.

To make certain parts better understandable, we have to give you some tools and products on hand. Therefore, we are in favor of highlighting open source over commercial solutions. We don't want to give you any bias, so always be aware that for everything we mention, there are at least a handful of options available. Some tools we might use as a reference are chosen because we know them better or because we see them more often used in the community than others.

Creating architecture should never include only one perspective. As with every previous topic, we have to consider the different angles from which stakeholders look at a platform. I don't want to give you a full set of the traditional enterprise architecture management tools on hand, but I would like to provide you with certain relevant visualizations and tools.

Platform component model

Stephan Schneider and *Mike Gatto* introduced the first version of an IDP reference architecture model in 2023 *[5]*, which the community reworked multiple times. Unfortunately, this model lacks some important perspectives, either because of the naming or simply because a so-called plane is missing. Therefore, we developed a more holistic approach to the model, which includes the following planes:

- **The developer experience plane**: This covers all first-row components needed for software development and infrastructure engineering
- **The automation and orchestration plane**: This clusters the tools to build, test, deploy, store, and orchestrate applications and infrastructure
- **The observability plane**: This includes any tool needed to provide transparency and visibility of what's happening in the platform and its apps
- **The security and identity plane**: This plane contains security-enforcing and supervising systems, user and account management solutions, and secrets and certificate handling tools
- **The resource plane**: This is a container for all computational resources that can be utilized for a platform
- **The capability plane**: The plane within or on top of the platform, such as Kubernetes, that provides services for the users

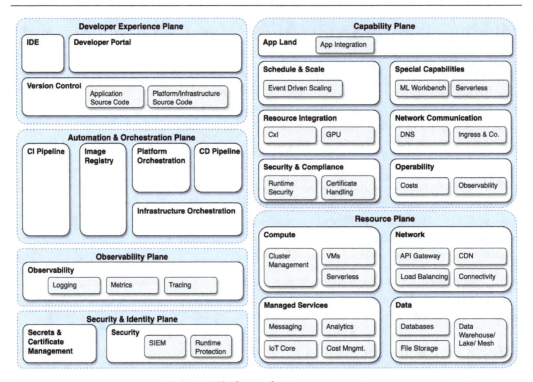

Figure 2.4: Platform reference components

This perspective on the components shows that a platform is definitely more than a single server on which I just install some container orchestration or developer portal. Often, those components have multiple places to live. This is a kind of link, looping the relevant elements together. For example, to enforce security, we will find relevant elements in the resource plane to adhere to compliance and deliver hardened configurations, such as external systems to collect data and analyze them, on top of the capability plane within Kubernetes to scan containers and protect networks, but also in the CI/CD to deliver only hardened containers. You can see that a single topic can live anywhere within a platform.

Platform composability

A platform consists of many parts that are constantly evolving. Alongside this development, your users and organization might also evolve. A huge problem we frequently discover is the reluctance to replace some parts of a platform. While there are often arguments about priority or budget, the most common problem is that it can't be replaced easily. Sometimes, this is because the platform and component are old and literally screwed into each other, and sometimes it is because the technology choice wasn't good and introduces a chain of dependencies that become hard to replace. But there is also a third reason: forcing your own way of doing things into the wrong tool.

Composability is a principle, maybe an ideology, and definitely an approach and decision-making that platforms should be built on. Components are composable when they interact with each other without the need to develop something in between for translation, such as scripts or serverless functions. Composable components can be replaced with the least effort possible and with the smallest to no impact on the platform functionality. Kubernetes is an environment that drastically supports this approach and becomes, therefore, the de facto standard to create your platform. However, the components of a platform are distributed, and not all of them run on Kubernetes. This makes it a challenge to find the right tools.

Composability also helps in the right adjustment of a platform. A generic, one-size-fits-all platform always sounds good on paper, but it is unrealistic and increases complexity. The more components you have to bring into a platform, the more likely it becomes that they don't play together well. Therefore, being able to provide adjustments for certain use cases without changing the entire architecture will increase its useability. Another strong benefit of composability is the separation of concerns.

In *Chapter 1*, we highlighted the purpose of a platform as an abstraction layer for Plug and Play specialized solutions. Strictly enforcing composability will cause a lot of work on the organizational level, but it will allow specialized teams to provide their core capabilities right into your platform.

To evaluate if a component for your platform is composable, you have to look for four abilities:

- **Orchestrateable**: It can run in a modern dynamic environment, such as Kubernetes, without being impacted
- **Modular**: It doesn't require any other tools or has technical pre-conditions, but can utilize those to extend its capabilities
- **Discoverable**: It needs to interact with other components following standardized APIs, or following concepts such as the custom resource definition of Kubernetes
- **Autonomous**: It must be independent from other extern inputs and operates by itself

Now, we know what to look for when designing the platform and choosing platform components. In the next section, you will learn about the things between those components that keep the platform together.

Dependencies and the hidden glue

No matter how composable the platform components are, there will always be dependencies and some *hidden glue*. This hidden glue is the information that flows between the components and makes them react to each other. As architects, it is our job to ensure that the information is not contradictory. This information is important but not well understood.

Managing and handling dependencies

First, let us build on the composability by visualizing dependencies between components. For this, we can use a simple dependency matrix. This tool can help you make decisions and plan future components. Also, it requires some maintenance effort; many platforms depend on the knowledge of single engineers and, as the platform becomes more complex over time, documentation does too. Mapping dependencies is therefore the shortest way to transparency.

In the following example, we have **tools A-C**, such as **container network interfaces** (**CNI**) **1-3**. Now, those tools can be any kind of tool from the reference components list we talked about before.

In the dependencies, I prefer to be explicit about which kind of dependencies we have there. Therefore, you can use things such as the following:

- **Unidirectional and bi-directional dependencies**: You can use these to describe whether tool B depends on tool A or tool C depends on tool B as tool B depends on tool C.

- **Enhancing dependencies**: Enhancing dependencies unlocks more capabilities and features if established, but it doesn't have to be. For example, if you have a tool with manual user management, now you can utilize **single sign-on** (**SSO**) for it, which reduces your manual efforts.

- **Conflicting dependencies**: As the name suggests, those two things don't play well together.

Legend: u = unidirectional, b = bi-directional, e = enhancing, c = conflicting

Y depend on X	Tool A	Tool B	Tool C	CNI 1	CNI 2	CNI 3
Tool A					b	
Tool B	u					u
Tool C		b				
CNI 1			e			
CNI 2	b			e		c
CNI 3		u			c	

Figure 2.5: Dependency matrix

Have you seen it? In the example of bi-directional dependencies, the relation between tools B and C is not correctly shown. In the column of tool C, the bi-directional market for tool B is missing. When you implement your dependency matrix, you need to give clear guidance on how to track those relationships. I think it is most useful to set the marks for both tools. However, then you have to be clear about the reading direction or you have to introduce more markers. For example, you could use *(u)* to define an incoming dependency.

Dependency management is a continuous task. Over time, you run the risk of forgetting to add new tools. Therefore, whatever method you use to implement new features, it might be a good idea to create an automatic sub-task to update the dependency matrix, too. A big benefit is also to evaluate this matrix from time to time, which might identify potential bottlenecks or problems you can work on.

Applications x platform – a reciprocal influence

The **reciprocal influence** is defined as related to each other in such a way that one completes the other or is the equal of the other. You have a certain kind of mutual influence in the relationship between the platform and your users' applications. For example, an application needs to scale drastically. As a platform, you support this behavior but also guarantee that it doesn't have an impact on other applications. On the other hand, the application decided to go with your platform because you support its demand. A platform without a user is useless. To perform best as an application, you should run on that platform.

However, this example is very generic. Reciprocal influence also affects elements outside the layers defined in YAML files. It is also the demand and requirements it causes without asking for it; it is the team dynamics between both implementing parties; it is also the effect one solution has on others in the way they work.

Therefore, *dependency* would be the wrong word to describe this relationship. It is a coexistence, a collaboration, pulling together in the same direction. For a platform engineering team, and you as an architect especially, you have to keep this reciprocal influence in mind when you design your solution. To make this your own superpower, you have to educate the platform engineers and users about these effects and focus on how they can get the best out of the platform without feeling any restrictions.

> Reciprocal influence
>
> The reciprocal influence is the invisible play between the application and the platform. Understand it as a joint force that makes the platform and the users' application better together than they would be without each other.

Another place where such reciprocal influence becomes visible is in processes.

Opportunities of process definitions for a platform

Platform engineering lends many approaches from the DevOps methodology. In a DevOps role, some parts focus on automation. Those automated steps translate into defining processes. So, we also could say that platform engineers create technical processes for them to be implemented as a platform as a product to make it function in a user-oriented way. However, those procedures are not very common in explicit design decisions. The form of the process (or how the process looks like) depends on the way certain components work. With respect to platform composability, this might be a given approach, but could we improve the platform by explicitly designing the processes?

It shouldn't be turned into a business process modeling. Too many advanced features that are backed into tools such as GitOps or operators on Kubernetes can have multiple variables based on decision paths. It would be hilarious to draw all of those. However, it is helpful for the process to be documented and designed to understand the platform. The difficulty is in the multiple entry points processes, which can start at the following:

- Infrastructure provisioning
- User onboarding
- Application deployment
- Incident handling

Each requires its own process definition.

You have the opportunity to find optimization potentials with a focus on user-friendliness and handling. Leveraging the process perspective differentiates user-centricity in your platform development from the more common, pure, tech-focused definition of features. Unfortunately, we can't give you more insight at this point in time as this discipline is still fresh, and there is barely any knowledge sharing happening around it. In *Chapter 3*, we will dig deeper into this topic.

The vendor lock-in discussion

To conclude the topic of dependencies, we have to touch on the discussion on vendor lock-in. It is part of strategic and architectural decisions and often pushes for strange outcomes. At this point, you should be fully aware that you will always have dependencies, and a vendor lock-in is nothing more than that. As open source brings a fantastic foundation to avoid proprietary software, it is also used as an argument to build things by yourself. The important question to ask is this: With which speed and service quality do you want to provide the platform to your users? *What are your expected operational costs?* In addition to that, you have to consider aspects of the available team, skill sets, and which costs your organization prefers (**operational expenses (OPEX)** or **capitalized expenses (CAPEX)**).

Some different perspectives on vendor lock-in are as follows:

- **Better build it yourself**: Building certain capabilities entirely by yourself might feel like your ultimate freedom and avoid any kind of vendors. The truth is you just become your own vendor. At some point, you don't want or can't invest anymore in your development (common project behavior). With that, you as your own vendor retire from your own market and everything that you were afraid of becomes true.

- **Picking best in class per problem, not per context**: In the range of observability, a large number of options exist for monitoring, logging, and tracing. A common approach to this is to pick the open source tool that fits the best reach case. Around this, you then have to implement a log collector and agents to forward metrics. What you have achieved is a fully composable observability solution. On the other hand, you are losing in comparison to a single vendor solution capabilities, such as context awareness, scalability, or operability. I don't want to say proprietary software is better; however, I have seen organizations that use 10-14 different tools, in different versions, running on the largest scale and having high operational costs far too often. In fact, for a toolset, that practice isn't considered helpful when it comes to incidents. In this case, we unfortunately tend to avoid looking at the truth and finding excuses for why it is this way better.

- **Focusing on the wrong things**: People still focus on the tasks of deploying Kubernetes. It's considered a challenge, and it becomes a never-ending story. Consider utilizing tools, opinionated platforms, or commercial solutions to get your container orchestration done, and focus on the value-generating part of your platform.

Vendor lock-in is an emotional discussion point on the agenda of an architect defining a target. You have to keep your eyes and mind open and truly consider what is important and what is just there to feed the ego. Complex and unmanageable solutions are nothing that will make one win awards for them.

Until now, you have learned about dependencies and how to handle them. In the upcoming section, we will learn how to make them visible.

Reference architectures

Reference architectures are valuable assets for a platform engineering team, your platform, and interested users. They build the foundation for discussions and decision processes and help to define future work with some foresight. Unfortunately, many platform teams can't provide a clear picture of their system.

Provider-specific reference architectures

Now, those reference architectures might change and look differently for different providers you might use. The platform engineering community [6] has collected a couple of options and templates to build on. A good starting point is the single cloud provider template, here with a flavor of AWS. However, the capability plane is still missing. The focus here is on the cloud provider as it often defines a huge amount of components for your platform. As you can see in the next figure, the **cloud service provider** (**CSP**) services can be found at any part of the architecture. You could draw an even more opinionated target utilizing a CSP CI/CD, secrets management, and cloud IDPs.

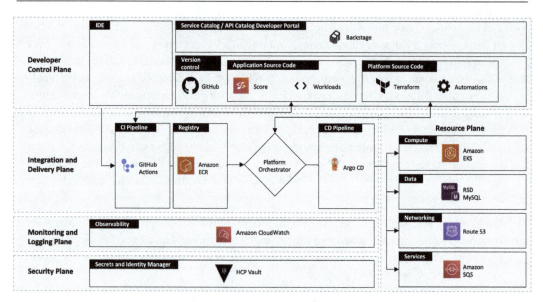

Figure 2.6: AWS reference architecture by platformengineering.org

In comparison to the single-provider platform, a multi-provider platform comes with its own complexity. Besides the resources used by each provider, we also have to consider where to store and distribute container images, how to provide observability, and how to manage secrets and certificates across the providers. You can see in the following reference architecture that it isn't simple to meet the expectations. With a multi-cloud platform strategy, you will grow a large number of issues that you have to take care of. Costs will drastically increase, implementations and platform behavior will slightly change with each CSP, and it will become difficult to define an approach that feels at all right. Aspects such as availability, security, scalability, and efficiency stand in conflict by being built on multiple environments.

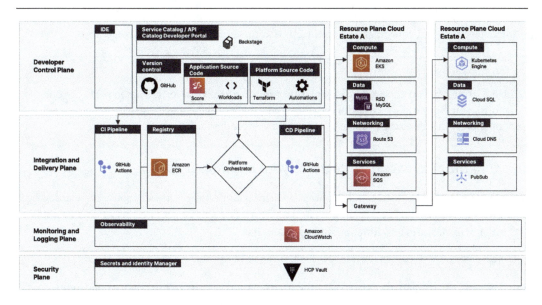

Figure 2.7: Multi-provider reference architecture by platformengineering.org

However, not every platform needs to run in the cloud. The number of options is endless. What we often see in mature enterprises are hybrid, on-premises, or cloud-extended environments that rely on established solutions such as VMware Tanzu or Red Hat OpenShift. In the following figure, we see how a platform would look with OpenShift as its core.

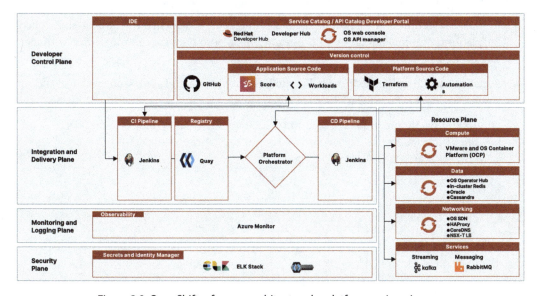

Figure 2.8: OpenShift reference architecture by platformengineering.org

It is important to adjust your reference architecture to the environments it will run within.

> **Important note**
>
> The environment you choose for your platform automatically prescribes some parts of your platform, whether we like it or not! Increasing the number of cloud and infrastructure providers exponentially increases the challenges for your platform design.

Extending the reference architectures with a capability plane

The weak spot of most reference architectures is their lack of a view of the capabilities provided by components running on the actual platform runtime. This perspective is required because users cannot use the platform's features and functionalities without it.

In the next figure, we have some references for the capability plane:

- **User Space**: The actual environment the user sees and can use to host their applications

- **Scale and Schedule**: Capabilities that support the user app in reacting to events, downscaling, or providing custom scheduling options

- **Network**: To handle the ingress and egress, automatically manage DNS entries or cluster load balancing

- **Heavy Babies**: Things you usually prefer to have managed rather than to operate by yourself, such as a framework for batch processing or message streaming

- **Resource Integration**: A vital component that bridges the actual infrastructure resources into the platform runtime; storage or GPU integration, but also resource control, such as with the tool Crossplane

- **Security and Compliance**: Provided capabilities to manage and store secrets, actively scan clusters and applications for CVEs and vulnerabilities

- **Observability**: This provides a simple and fast observability entry, unifies data collection, or helps to handle the costs

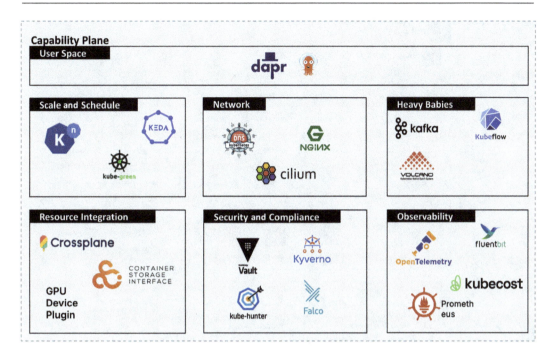

Figure 2.9: The capability plane with example tools

The capability plane is crucial for the actual application deployment and operation. It is the space where you have the greatest technical interaction, and if the user's app fails here, all other efforts invested are wasted.

A platform architecture goes beyond references

Recapitulating the reference architectures, they provide a simple-to-understand concept of the platform. What they don't provide is the actual architecture of the platform. Where is the source code versioning running? Where are the CI/CD pipeline components? Where is the container registry? How is the IDP hosted?

Therefore, we have to provide, as platform architects, an actual platform architecture that visualizes where those components are located. This gives deep insight into further dependencies, network communication and requirements, availability, and the scalability of the platform with all its components, or simply whether there are too many pieces built into it that actually do not help the platform's stability.

The following diagram shows what such an architecture could look like. You will also see that different components and network connections are spread around. However, when talking about a platform architecture, this is how it should technically be represented.

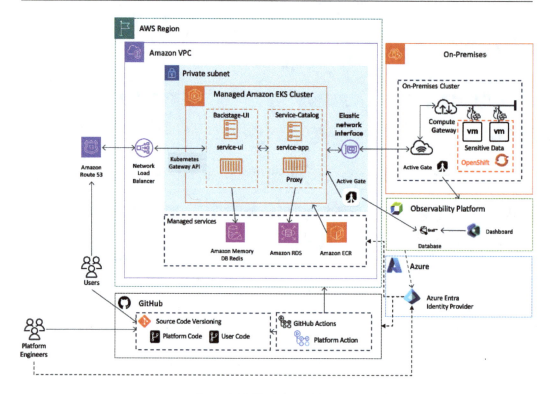

Figure 2.10: Platform architecture

It becomes difficult to find the right approach to show the infrastructure and platform components in one diagram. It isn't the cleanest approach, but it is necessary to highlight all dependencies and reciprocal influence. Keep it as simple as possible but as detailed as required. You might even split it into different diagrams to highlight certain pieces of information.

Owning your architectures

It seems like a lot of paperwork to manage and handle different versions and variations of an architecture. However, you might already have experienced that a platform changes faster than the diagrams. The important thing is that neither the diagram, as a result of your work as an architect, slows down the development of the platform nor does the platform development slow down the diagrams. I have seen projects where architects became an outstanding hierarchical institution. All of those projects failed and had difficulties in doing the right thing. As an architect, you are not all-knowing, but you are also not the platform engineering team. Working together, aligned, and in sync is the key, almost like the platform you want to build.

The same thing counts for your architectures. You have to own them but not rule them. An architecture must be a playground for ideas, a communicator for future developments, and a source of truth regarding the capabilities of the platform. That's just possible if you are responsible for driving it with the input of others.

Later in this book, we will cover a relevant topic when we talk about owning your architecture: dealing with technical debts.

Opinionated platforms and the cost of quality

While architecting a platform, two issues will arise: opinionated tools and the cost of quality. Both have very good arguments to be considered, but they also have very strong arguments against them.

Opinionated tools and products

Opinionated tools, products, and services emphasize a strict concept and don't provide alternatives. Practically everything comes with an opinion, but the question is how flexible and composable they are. A strong opinion on how something should be done makes the following decisions easier as the solution space shrinks. With a growing user base on such an approach, it might become dominant and eventually a de facto standard. We have to carefully consider them in our architecture and clarify an understanding of their potential future impact.

Now, the strict implementation of a concept can come with many benefits. For example, every solution is opinionated in the space of provisioning Kubernetes clusters. However, they provide a holistic solution that is consistent and usually ready for production. Such tools can also be faster in integration when they cover a large field of topics and pre-integrate other tools. This can also be, on the other hand, a downside, as if you like to implement other options, it will cause more effort to do so. Yet, even in some cases, it is almost impossible or way too expensive.

You should also differentiate between the visible and big concepts and the more invisible, hidden ones. The first ones are easy to identify and evaluate. They come with a large stack of pre-defined integrations and are usually based on a single vendor's technology. The small and hidden opinionated tools are more difficult, as they reveal their downside the more extensively you use them.

The cost of quality

There is this saying: *Better to buy expensive once than cheap often*. With platforms done right, this saying is wrong. Following a product mindset means a continuous investment. It is more like a garden you have to take care of day by day rather than a watch where you have a one-time investment.

The cost of quality is a multi-dimensional problem. A high-quality platform, most likely very opinionated, might be, for some use cases, the perfect fit. But the quality doesn't come for free and is based on high investments or operational fees. Within larger companies, you have something like internal cost sharing. That also means that you have to share the costs of quality. We frequently see in the end-user community that such platforms lead to frustration. The entry level due to costs is then too high for hosting simple applications, or too opinionated to run a very complex application system on it. This causes others to start building from scratch their own platform and solutions for their application. As you can see, this is a strange problem. Due to the high costs of the platform, the operational costs for a user become high too, while the opinionated solution doesn't actually give what one needs. That is called the **cost of quality**.

However, you also can't go around this issue by plugging and playing some random tools fast together to provide a cheap and as open as possible platform too. This might lead to an initial fast ramp-up of projects to adopt the platform, but I can promise this way, the faster they come to you, the faster they will leave your solution. Remember, with a product mindset, you want to enable your users to do things by themself, while your platform takes away all heavy lifts. Ideally, your user also doesn't need to become an expert in all the technologies used.

On a strategic level, the problem of the cost of quality affects many decisions. We have seen platforms that went through years of development only to be challenged in the end by their performance/price ratio.

You will need to find the golden path for your platform. However, don't assume that a perfect platform matches what others need. If you phase this issue, it is a clear sign that you have lost track of a product mindset with a user orientation.

Creating your own architecture

It's time for action. So far, we have covered all relevant input sources, topics of concern, and perspectives to consider so that we can start putting together an architecture. Therefore, we prepared a template that you can find in the GitHub repository for this book or as a `Miro` template for collaborative working.

Within those templates, you will be guided through different steps and architecture diagrams to build your platform architecture:

1. Create your reference architecture.
2. Focus on the capability plane and components.
3. Define the infrastructure architecture for the platform.
4. Visualize the control flow of the platform.

As architecture is a living artifact, you have to rework it repeatedly, following the *PLAN, DO, CHECK, ACT* cycle.

Take your time going through the templates:

- **GitHub**: `https://github.com/PacktPublishing/Platform-Engineering-for-Architects/tree/main/Chapter02`
- **Miro**: `https://miro.com/miroverse/platform-architecture-workshop`

Creating your reference architecture

The reference architecture outlines your planned components, categorizes them, and anchors them at the forefront of your platform implementation. This is also your source of help in defining the dependencies map at some point.

To fill your reference architecture diagram with some life, it is good practice to either start with the already fixed components or the given infrastructure. Fixed components can be, for example, CI/CD tools or security solutions. To start, the infrastructure becomes relevant as it influences the resources provided. If you go on with a public cloud provider, many services can be used as a managed solution. In some cases, this also defines which **Infrastructure as Code (IaC)** you are going to use, or the other way around, if you take just an infrastructure-as-a-service provider, it means you have to think about potential services that are running somewhere else in the platform. Most likely, they will end up in the capability plane, which then causes this question: Do we need a dedicated cluster for shared services or do we implement an approach that the service will be deployed on site of the users' namespace?

Next, take on the automation and orchestration plane. For some reason, this plane is relatively pre-decided and provides a solid ground for decisions for the developer experience plane. You should be sure that the given solutions are on a newer release and can integrate with other cloud-native services without any scripting.

Last but not least, you have to come up with the observability and the security plane. As said, it can be that you have already defined solutions in place. But especially for organizations where this platform engineering and cloud-native journey is new, you might challenge the given solutions. Security tools need to be able to understand the cloud and Kubernetes. They need to be integrated into those platforms and be able to handle their very dynamic nature. The same reason counts for observability and operational solutions. You don't want to get an alert every time pods get killed or moved around in the cluster.

Most importantly, you bring all the perspectives together in the end: the platform principles and purpose, its key users, composability, and the reciprocal influence between all of those moving components. But your design doesn't need to be perfect. Before you go too deep into the design process, read on, as we cover the topic of the TVP later in this chapter.

Focusing on the capability plane and components

When the reference architecture is standing, we have to extend it in the next step with the capability plane. Certainly, this is a big step in itself, because there are literally hundreds of open source solutions available, for any demand within the capability plane. In case you have never seen the CNCF landscape, the following snapshot shows around a third of the total available CNCF-listed open source projects [7].

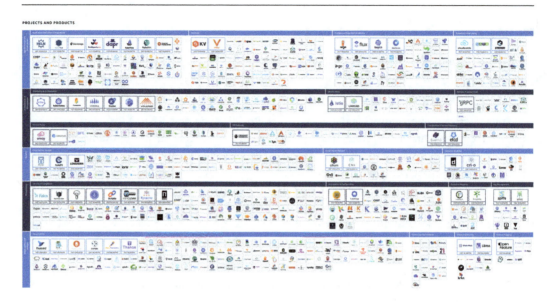

Figure 2.11: CNCF landscape (The image is intended as a visual reference; the textual information isn't essential.)

Be aware that there are at least a handful of other open source organizations that also manage dozens of projects, which you can count within the cloud-native space.

The good and the bad thing about the cloud native universe is that for every micro-problem, you have at least three ± 1 options to solve it.

For the capability plane, it is also recommended to work from the bottom up:

- Which integrations into the resources do we need?

- How do we provide the operational data to the observability solution? Do we need to split between infrastructure and application space, and if so, how?

- Ensure that security and compliance measures are correctly transported into the user's land.

- How do we handle network traffic? Do we need an encryption within the cluster?

- Do we expose an API, or do we integrate it with a cloud API?

- What might be required for scaling and scheduling? Are there additional features that need to be provided?

Expect to work in the capability plane most of the time. With every new user and product, new requirements will come up. The best part is when you can observe how the products are evolving and demand more advanced features.

Again, focus first on the essentials. Getting your network management and storage integration right is more important than providing a machine learning workbench solution or new container runtimes, such as **WebAssembly (Wasm)** *[8]*.

Defining the infrastructure architecture for the platform

After the fun part of defining the reference architecture, we have to go straight into the serious architecture definition. What sounds like a joke is fairly often skipped, yet many platforms can't provide this view. In fact, this diagram is more relevant than the reference architecture for the implementation, architectural discussion, and security and compliance checks. However, it is also more complicated to understand.

As explained previously, we have to combine on one side an infrastructure architecture with the respective components of the platform. This helps us understand where everything in our system's landscape is located and which connections are required.

The diagram has to highlight the different environments in which computing infrastructure is used, which managed services are consumed, how those are connected to each other, and which external connections exist. Every organization has its own way of drawing such diagrams, following given standards, or being more simplified. Our template tries to give a middle ground.

You might also vary with the highlights on the architectural diagram. We often see huge relevance in visualizing the exact network connection, network isolation, or protocols. The combination of an isolated environment with a corresponding control flow and how updates are played into it will utilize both types of diagrams to maximize understanding.

Remember to include things like management environments/clusters, staging environments for integration testing, or even one step before the development environment in the diagram. For continuity purposes, backup and recovery might be helpful too. You see, the more you work on it, the more details will float into it. When a diagram becomes too complex, you should provide different versions.

Visualizing the control flow of the platform

What might be new to you is the **control flow**. It is the first good step of a process-like approach to get clearance on certain relevant paths to manage the platform and user deliverables. There are multiple flows within a platform; these three are the most prominent:

- **Infrastructure provisioning**: This flow visualizes the importance of the infrastructure test environments and how changes on that level can impact them, from committing the IaC manifest in a versioning system through deployment, testing, and final rollout.

- **Code to deployment**: This is most relevant for the platform users. How does one's committed code get picked up to be built, integrated, scanned, and tested? What does the development and life-setting procedure look like?

- **Platform capability management**: Especially relevant for the platform engineers taking care of the cluster-delivered components.

There are many other smaller control flows that can help us understand the platform, such as secrets and certificate management or user management.

Moving on from the paperwork to real use cases, we will have a look at some examples of different platform flavors and implementations. In the next part, you will extend your perspective from generalistic platforms to specialized solutions.

Exploring platform as a product – use cases and implementations

We talked a lot about the platform's purpose, such as *self-service first*. But how do we achieve this and what use cases do we want to provide as a self-service?

Every organization has slightly different use cases and implementations for a platform. That's because every organization has a different technology stack and a different legacy/existing set of tools and processes that are candidates for optimization.

In this section, we discuss use cases we have seen in organizations we have worked with over the years. We can assume that some of those will be good candidates for your own future platform engineering initiative.

Finding the experts and the bottlenecks they cause

There is a great blog post from *Thoughtworks* with a quote that allows us to get started with our search for good use cases *[9]*.

> **What makes something a platform?**
> Platforms are a means of centralizing expertise whilst decentralizing innovation to the customer or user.

In other words, in any IT organization that is producing, shipping, and operating software to support its business, we can find experts on topics such as provisioning infrastructure, access control, building and deploying artifacts, observability, quality engineering, release management, capacity planning, **site reliability engineering** (**SRE**), incident response, data analytics, automation, business insights, and so on.

The goal of our platform is to identify use cases and which expertise they require, and then provide those use cases in a simple way as a self-service for everyone to consume without needing to go to the expert every time.

> **Centralize expertise through a self-service**
> Allow everyone to get their job done without having to be an expert or ask an expert in all the areas needed to get their job done!

If we now look at the full **software development life cycle** (**SDLC**) within those organizations and list all the tasks, experts, and time needed from the initial idea of a new app or feature until it is released to end users, we will be able to identify a lot of bottlenecks that slow the SDLC down. This is either because certain tasks are manual, or because they require somebody with expert knowledge to get a task done. As experts are typically rare, they become shared resources that teams will have to wait for when they want to get their new feature out the door.

In *Chapter 3*, in the *Understanding the existing SDLC* section, we will have a closer look at how to best understand the current SDLC or the *Value Creation Journey* within an organization. You will learn about different approaches to understanding the life cycle of an artifact, the tasks involved, the dependencies, and how to track time to identify good candidates for automating those as part of a platform offering.

Let us now explore some of those use cases and implementation options from organizations we have worked with in the past!

Centralizing expertise as a self-service use case

While there are many more of those use cases, here are a couple of examples that you will likely find yourself implementing as a self-service feature as part of your platform.

Provisioning compliant environments

Depending on the industry, there are certain regulations on how data has to be stored, how environments have to be secured, and what type of reporting needs to happen. To make sure that every environment that is requested by any team in an organization is compliant with all those rules, we can implement this as a self-service capability of our platform.

Here is a user story for that use case: "*As a data scientist, I want to validate my new data models against a production-like dataset!*"

There are several details that have to be worked out in order to break this story into its individual tasks, such as the following: Where do these data sets come from? Does the data scientist want to choose one from multiple available ones? What's the output of the validation, where is it stored, and who has access to it? How do you select the data model? How long does this validation run and do we need a max time to shut down the environment to avoid unnecessary costs? What type of auditing data do we need to produce as this is about accessing production data?

Another similar user story could be this: "*As a QA engineer, I want to run my manual tests against the latest version of our software connected to a production database, using a browser/OS combination that matches what 80% of our end users use!*"

Similar to the previous user story, we need to ask some additional questions, such as the following: What is the software product you want to test and how can it automatically be deployed? Where can we find the data about the browser/OS usage of our production users? How long does this environment need to be available? Can we set an expiration time for that provisioned environment? What data is the engineer allowed to see and which data are they not allowed to see, as we are dealing with having access to production data?

The implementation of such a use case can vary widely but we should always start and think about the *end user journey*. We probably want our data scientist or quality engineer to log on to our internal development platform portal. There, they can select the use case of *provisioning a compliant environment*. Then, they can fill out the relevant data to answer all those questions that came up. That input can then be used to create that environment and make it available to the team that requested it.

Running performance and resiliency tests

Performance and resiliency testing should be part of every release of software to ensure that new features do not impact the UX by becoming super slow. We also want to make sure that our new software is resilient against unforeseen scenarios, such as a network connectivity issue, a slow or unavailable backend service or data, or an unexpectedly high load.

While there are many tools that can generate traffic (load testing tools) or simulate problems (chaos engineering), those tools and the environments they need often take a lot of expertise to set up, configure, run, and later analyze the results.

Instead of having our performance, site reliability, or chaos engineers become the bottleneck, we can strive to centralize that expertise and provide this as a self-service to our engineering teams. Here would be the right user story: "*As a development team, we want to know whether the latest version of our software has any performance or resiliency degradations!*"

There could be multiple iterations of this self-service capability that we could implement. Starting with provisioning the environment that contains the relevant tools, all the developers need to do is execute the tests and wait for the results. Ideally, though, we want to end up in a situation where this can be fully automated and even integrated into our development process. We should aim for something like this: "*As a development team, we want to get a performance and resiliency indicator on major Pull Requests so we know whether the latest code change is good enough to be promoted to production!*"

Like in the previous example, we could start with a portal where teams can request the performance testing environment. To fulfill the second, fully automated user story, we need to think about providing an API that can be called from the CI/CD pipeline system that gets all relevant input parameters and then returns the actual result of the executed test.

Onboarding of a new application

Creating new applications or services is what development teams do. To do so, they typically have to go through a lot of different steps, such as creating a new Git repository, adding boilerplate code and metadata settings, configuring the build (CI) pipelines, and many more steps.

If every development team always started from scratch, we would not only end up with multiple different ways of doing essentially the same thing, but we would also have a lot of duplicated work across all development teams that take the time away from actual coding.

A good user story for a platform self-service feature would be this: "*As a development team, we want to create a new application based on a fully configured template so that we can focus on writing code and not deal with how we build, deploy, and operate!*"

Looking at the full SDLC, our self-service can even start a step earlier, with the feature requirement that was created by the product team with tools such as **Jira**. The following shows the end-to-end onboarding workflow of a new application in one of the organizations the authors have worked with:

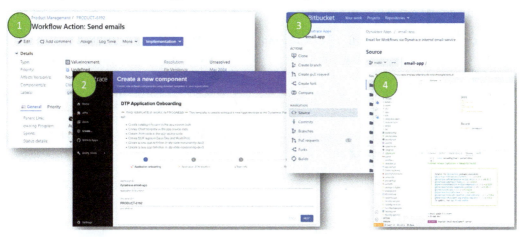

Figure 2.12: End-to-end application onboarding as a self-service (The image is intended as a visual reference; the textual information isn't essential.)

For development teams, the journey starts on the Jira ticket. They then go to Backstage, the IDP chosen by this organization, where they walk through the self-service *Application Onboarding* wizard. That wizard creates a new Git repository that is pre-loaded with ready-to-go code templates, pipelines, deployment instructions, observability configuration, ownership, and more. Once code gets committed, the pipelines automatically deploy the artifact in a development code space that gets automatically connected to Visual Studio Code!

Access to observability data for incident response

Once new software gets deployed into production, the focus for developers typically shifts to the next application or feature. When disaster strikes, though, it's time for them to support the operations or SRE team with troubleshooting to fix any issue that came up.

In many organizations, access to telemetry or observable data in production is limited to selected teams that take care of those production environments. There are also compliance reasons as not everyone should have access to potentially sensitive data. But when sitting in the so-called *War*

Room, it's important to quickly access the relevant observability data without having to fill out too many request forms, copying things from one environment to the other, or converting data from the production observability tool to what developers are used to.

In *Chapter 3*, we walk through this use case in more detail so we'll keep it to the user story here: "*As a development team, we want to get easy access to the relevant observability data for a production incident to solve the problem fast!*"

Now that we have a couple of use cases, let's see how we can translate those ideas into something our users can actually use and benefit from.

Understanding TVPs

In product management, we typically talk about a **minimum viable product** (**MVP**). The MVP defines the simplest version of a product that you need to build to sell it to a market. The concept of the MVP was initially introduced through the *Lean Startup* movement driven by *Eric Ries [10]*. I encourage everyone to either read the book or the excellent blog posts on the `theleanstartup.com` website.

The *minimum* in MVP refers to a version of a product that helps the start-up team validate the hypothesis that their idea really solves a problem or pain. All of those Lean Startup ideas can directly be applied to platform engineering like this:

- Our team is the start-up
- Our platform is the product
- Our target market is our internal users
- Our hypothesis is that it delivers the purpose

In Lean Startup, the MVP is also defined as the first version of a product that is *good enough* to ship. It is good enough because it already solves problems, while it might not be feature-complete. It is good enough to get it into the hands of users, validate our hypothesis, and get early feedback on which we can iterate and make the next version better.

Now, let's dive deeper into how we can apply all those Lean Startup ideas and define our own MVP – or, as we like to call it, our TVP!

Finding your TVP use case

In a previous section, we already explored which use cases we could provide as a self-service through our platform. We identified them by looking into where experts are currently needed to get a job done or where there is a lot of manual work involved.

As with any new software product, there are probably a lot of great features we want to implement. The question is this: Which is the one we should start with? To answer this, it's best to prioritize based on several dimensions. One such approach is the *ICE scoring* model but feel free to use any scoring model that you are familiar with or already use in your organization *[11]*. **ICE** stands for **impact** (the impact we will have with this feature), **confidence** (how confident we are that we really achieve the desired impact), and **ease** (the level of effort that goes into implementing this). We give every indicator a value between 1-10 and then multiply the numbers. If we do this for every feature idea, we can easily compare and come up with a first prioritized list. Here is how this could look like for the four use cases we discussed earlier using some fictitious numbers for the ICE scoring:

Use Case	Impact	Confidence	Ease	Score
Provision Compliant Environments	6 (good impact)	5 (middle confidence)	2 (not easy)	60
Running Performance and Resilience Tests	6	5	5	150
Onboarding of a new application	8	7	3	168
Access to Observability Data	8	8	3	172

Table 2.1: ICE scoring example

As we said, the ICE scores in the preceding table are just fictitious. However, the examples should give you an indication of how this scoring works. Another form of it includes reach as well, so that is called RICE. **RICE** is simply multiplying reach, impact, and confidence and dividing the result by effort – giving us an easy-to-use score.

Based on the preceding example, we could argue to either start with *Onboarding of a New Application* or *Access to Observability Data* as they scored highest. On the other hand, we could also argue that *Running Performance and Resiliency Tests* would be a better candidate as it seems to be the easiest to implement.

Whether you use ICE, RICE, or any other model, it should help you make a decision on which use case to tackle first with the goal of quickly making an impact. Before we start implementing, let's talk about how good that first implementation has to be!

Good enough versus perfectly done!

We probably all agree that a product is hardly ever perfect or done! There is always the one or the other feature that we would like to see or that annoying bug that we hope somebody can finally fix. When starting with a new product, such as our platform, we need to change that thinking and be good with a first release that is *good enough*. This doesn't mean we are taking shortcuts; it just means we need to set aside our perfectionist mentality and have the courage to say, "*It's good enough – let's ship it!*"

In the Lean Startup, there is a reference to the launch of Google Maps *[12]*. Seems that the team was demoing their new dynamic web (using AJAX) map solution to the Google senior management team. Their approach was the first of its kind. The management team was very impressed, even though the development team still considered it just an early prototype. Larry and Sergey, so the rumor goes, simply said, "*It is already good enough. Ship it.*" While the development team had reservations and fears, they went ahead as asked. The rest – as we all know – is history: Google Maps was and is still a huge success. This success was driven by the fact that the solution just did one thing – but that thing was done extremely well and was a key differentiator to everything else out there. Shipping it just with a limited feature set was what caught their competition by surprise and gave them a leadway.

How does this translate to our TVP? Unlike the Google example, we do not have to fear competition – or do we? We actually do: our competition is the development teams that either waste their time by doing things the same way or starting their own initiative to build tooling and automation to solve the problem just for them instead of thinking about how to do this at scale for the whole organization.

That means that we must not be perfect to deliver the first implementation of our TVP, but it must be good enough to help us show that we provide value. What that value is is what we need to specify in our hypothesis for the use cases we implement.

TVP – validating our hypothesis

So, we picked our use case and we know that our first delivery has to be good enough for our end users to use it and get value out of it. But what is that value? How can we measure and prove that our platform capability is actually having an impact?

This is where our product hypothesis comes in. Looking at the same use cases from previously, we could come up with the following assumptions on the impact we want to have:

Use Case	Hypothesis
Provision Compliant Environments	80% fewer compliance violations when validating new data models
Running Performance and Resilience Tests	50% fewer scalability issues in production as we identify and fix those problems earlier
Onboarding of a new application	20% reduction of lead time of a new application
Access to Observability Data	50% reduction of troubleshooting time in production issues

Table 2.2: Product hypothesis to validate use cases

The hypothesis is also a great way to pitch the idea to those people within the organization who need to provide the funding for our efforts. In the end, we are selling internally – and we need to make a strong case for why we would invest time and money in building a new internal development platform. Those value statements that we call hypothesis will most likely resonate really well with your leadership.

The last question that remains is this: How can we measure and validate our hypothesis? For some, it should be easy to assume we have things such as *number of tickets for compliance violations*, *tickets for scalability-related product issues*, or *development time booked on incident response tickets*. The trickier ones will be around lead time, as how an organization defines lead time first needs to be explained. Does it start with the creation of the initial feature request or when the developer started to work on it? Also, how do we measure the full end-to-end stream? While all of this is possible, we must make sure we know how to measure the status quo to then be able to compare it to the numbers once we have our TVP in place to validate our hypothesis.

Build, measure, and learn

We know what we want to build, what our hypothesis is, and how we measure it. Now, it's time to put this into motion. Like with any agile product development, we want to build, measure, and learn. We want to involve our end users as soon as possible. The best way to do so is to already include and learn from them when we are still in the prototyping phase. Continuous feedback helps us make important decisions early instead of waiting until we have a final version that our users reject because we missed something very obvious.

The process can best be visualized with the following diagram that you also find in lots of literature inspired by the Lean Startup movement:

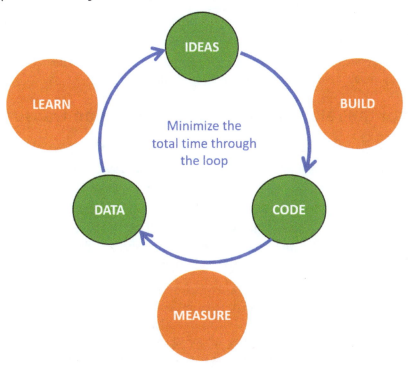

Figure 2.13: Build, measure, and learn to get to your TVP

It is a continuous loop that tries to minimize the time between the build, measure, and learn stages so that we can course correct or pivot in case the feedback from our measurement indicates that we are off the wrong path.

A couple of milestones that are good to consider in that loop are as follows:

- **First prototype**: Show this to interested and potential individual power users. They will give you great early feedback.

- **Good enough**: Once you reach a stage that delivers on the promise of the user story, expand it to a group of early adopters to get broader feedback.

- **Hypothesis achieved**: Once you proved the hypothesis with your first early adopters, it's time to open and promote this to the rest of the org. Use your early adopters to promote the new capabilities for you!

Here we are. We have our TVP!

> **The TVP**
>
> The TVP is that version of our platform that provides value to our end users. It allows the platform engineering team to validate our hypothesis that it fulfills the purpose of our use cases (e.g., reducing cognitive load). It allows us to collect the maximum amount of validated learning with the least effort for our next product iteration.

When you start introducing a platform, it's important to track its adoption and identify whether you are on the right path. Next, we will help you define KPIs to do so.

Looking at the relevant KPIs to make adoption transparent

Do good things and talk about it. (Georg Volkmar Earl of Zedtwitz-Arnim, 1978.)

A common issue we see across many platform implementations is the lack of evidence of its benefits. In theory, that all makes sense, but let's put some numbers behind it. In addition, they help to understand if you are going in the right direction.

Therefore, it is helpful to define different **KPIs** based on which one can understand the progress of platform adoption. In the first step, you should reflect on what you measure already. Very often, we just need to combine available numbers in the right way to be able to define new performance indicators.

First, we need to understand the metrics, logs, and traces, define how we get the values they represent, and what they mean. While all of them can be relevant sources, most will not be relevant to your platform's adoption. This just represents the system's side. Meanwhile, it should be clear that we are more interested in the user and their experience. Support channels, ticket systems, open pull requests, the number of onboarding requests, and other user interactive tools will be helpful in understanding the adaptation. The importance of those numbers will change over time. For example, the number of people requesting access to the platform will usually look like a hill with two flat sides. On the one side, because people are too shy to try new things, adoption starts slow. On the other side of the hill, fewer new projects will be onboarded to the platform. Another example could be the number of support requests. They usually grow with the number of users. However, as a platform targets self-service, those numbers can be expected to decrease to a minimum over time while your platform becomes better and users understand how to use it.

That's why, secondly, you have to define the current context of your platform's KPI. It might seem to be obvious that a mature platform might have different KPIs or that your stakeholders have to read those KPIs differently but believe me, it is not. It is on you to make this transparent, clear, and understandable to everyone.

> **Important note**
> KPIs and their current context of maturity need to be explicitly specified.

As an example, the following diagram highlights how the KPIs could look compared to an environment where a platform is poorly integrated. The numbers are based on a few measurements taken from different projects and generalized here. On the left side, there is no visible direction of improvement, or they keep developing up and down. The most obvious is an increase in support requests. Placing this in the context of fewer applications using it is a bad sign. Now, on the right side, we can see an improvement for any KPI. Most numbers will not go down to zero, but they will stay steady independently of platform growth, end-user growth, or other long-term impacts.

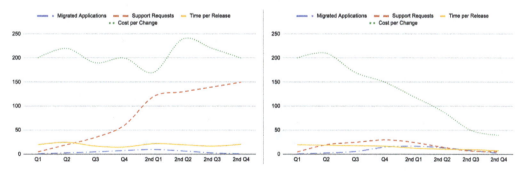

Figure 2.14: Comparing platform impact (left: no platform, right: with platform)

These numbers are missing context. When you declare a certain state of the platform through its life cycle, numbers become more meaningful. In the next figure, we added an example of segmentation.

With the first adoption support requests, costs per change are increasing, while only a few applications are being migrated. During the ramp-up phase, you can see that the number of new applications is increasing fast, and at the same time, the support requests go up. The cost per change stagnates. In the next segment of optimization, we can now see heavy changes. Cost per change and support requests are dropping steeply. The number of new applications is at its peak and slowly going down due to the fact that most applications that could use a platform have migrated.

Over the three phases, the time per release drops down. That's an effect of the automation and self-service and starts flattening out in the last segment. This is similar to almost all other KPIs that have found a steady development. In comparison to the other diagrams, we also changed the representation of the number of added applications. Now, you can see the number of migrated apps per quarter as well as the overall number of applications.

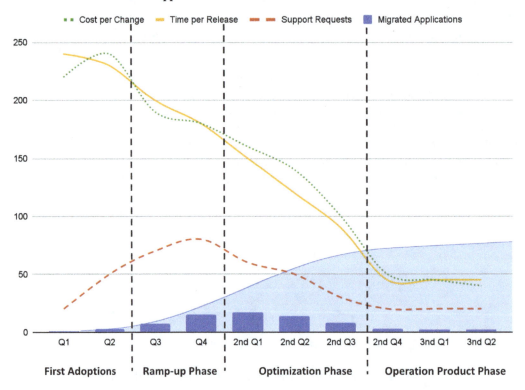

Figure 2.15: Example KPIs with the segmentation of the platform life cycle

Usually, when companies see such numbers, they would stop investing in it. With a product mindset, those numbers mean you are able to provide continuous improvements and innovation for a static set of costs, while customers are satisfied and enabled for fast development.

Defining platform adoption KPIs

We can use many technical numbers as a foundation for our KPIs. Those are easy to access and correlate. On the other hand, the interpretation of the numbers is up to you. The reduction of service requests can be because you have fantastic documentation, but also because no one uses the platform. As shown before, that's why it needs to provide a full picture. It might become complicated because every new KPI might add more details to the data. At some point, you will need to cut this.

For developers and DevOps teams, there are already a couple of frameworks that we can use as a foundation.

DORA metrics

DevOps Research and Assessment (DORA) is a start-up created by Gene Kim and Jez Humble, whom you might know from *The DevOps Handbook*. The DORA metrics are a set of four KPIs that can be used to measure the performance of a DevOps team and the impact of a platform on its users. Although they do not always fit well, they are widely adopted and used, for example, by systems such as GitLab *[13]*.

The DORA metrics are as follows:

- **Deployment frequency**: How often an organization successfully releases to production
- **Lead time for changes**: The amount of time it takes a commitment to get into production
- **Time to restore services**: The percentage of deployments causing a failure in production
- **Change failure rate**: How long it takes an organization to recover from a failure in production

I often saw that organizations change the term from *production* to *release* or *productive release*. That's because most companies have multiple stages a release goes through before it goes live in front of an end user. Due to every organization's specifics, there are multiple solutions to measure those numbers, but in the end, it's common to implement your own approach for it.

To measure the number of releases, you can run a simple API call against your favorite source code versioning system that supports you with your releases. These numbers can be dumped into a database and be visualized by a Grafana dashboard.

It is more complicated to identify the lead time. For this, you have to track commits and when they end up in a release. However, both metrics are simple to get and calculate. The change failure rate, on the other hand, becomes more problematic. For this, it is required that you include information about your releases throughout the whole deployment and production chain so that if a failure occurs, you can identify in the logs to which release this issue belongs. In addition, you then also have to qualify if it is really an issue caused by the release or not.

The time to restore service follows the same procedure. You need to identify when an incident occurs when it was closed successfully, and when it was resolved by a restore. Your incident management system can provide both of those metrics.

Now, what's so special about the DORA metrics is that they have researched thousands of companies to give an overview of how good or bad you are compared to others. From their last report in 2023, we can find the following categories and share of companies falling into those categories *[14]*.

	Elite	High	Medium	Low
Deployment frequency	On-demand	Between once a day and once a week	Between once a week and once per month	Between once per week and once per month
Change lead time	Less than one day	Between one day and one week	Between one week and one month	Between one week and one month
Change failure rate	5%	10%	15%	64%
Failed deployment recovery time	Less than one hour	Less than one day	Between one day and one week	Between one month and six months
% of respondents	18%	31%	33%	17%

Table 2.3: DORA metrics for different maturity levels

What is good about the DORA metrics is that most companies will be able to gather such numbers. On the other hand, we as platform engineers are missing some aspects around direct and personal feedback.

The following figure is an end-user Grafana dashboard that shows real-time data. In the top-left corner, you can see the number of deployed apps to production is 105 and the total deployments is 1.847. In the top-right corner, it shows the lead time for a change in hours, 612.5, the change failure rate of 0.522%, the deployment frequency of 5.02, and the MTTR of 199. The graphs at the bottom show the deployments per application.

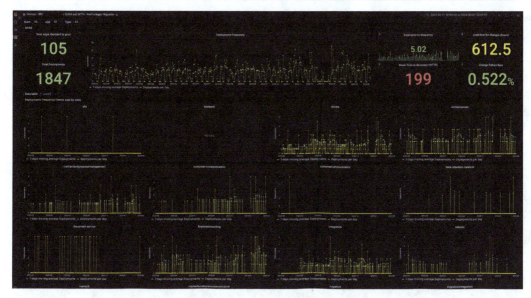

Figure 2.16: Real example of a DORA dashboard (The image is intended
as a visual reference; the textual information isn't essential.)

How can you get started with DORA? One way is to leverage **Keptn**, a CNCF Incubation project, which provides automated observability into application-aware deployments on Kubernetes. Keptn can be enabled on a namespace level via annotation and will create Prometheus metrics as well as OpenTelemetry traces to make deployments observable. Those metrics include the duration for deployments and deployment success and failure and provide dimensions for deployment, application, environment, and version. Out of the box, this gives you some of the core DORA metrics that you can put on your dashboard, as shown in the following figure:

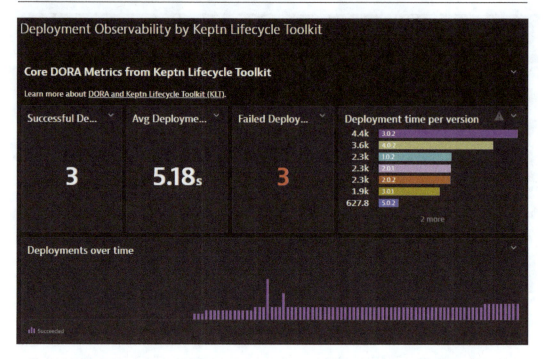

Figure 2.17: Keptn provides automated deployment observability to report some of the DORA metrics

Keptn provides additional capabilities, such as SLO-based deployment success validation, and also traces deployments across multiple stages to identify additional metrics, such as, "*Where do deployments stall in the promotion process?*"

To learn more, check out the Keptn website and DORA tutorials *[15]*.

SPACE framework

The **SPACE framework** combines five dimensions that assess a developer's team productivity. GitHub, Microsoft, and the University of Victoria developed this approach. It focuses on the way teams work together and their outcomes. Looking at the five dimensions will make this clear:

- **Satisfaction and well-being**: Collect information about overall fulfillment, happiness, and work-related mental health.

- **Performance**: Output-oriented metrics, such as tasks completed or releases published

- **Activity**: Focusing on activities such as the number of commits, merge requests, or code reviews

- **Communication and collaboration**: We will examine the teams' communication behavior and alignment, as well as usage review and commenting mechanisms

- **Efficiency and flow**: How smooth the development and deployment process go

You might get the feeling that you could use some or all of the DORA metrics to answer performance, efficiency, and maybe activity, and you are right. But there are more approaches you can take.

For satisfaction and well-being, we can use methods such as the **employee satisfaction score (ESS)**, the **employee net promoter score (eNPS)**, or the **employee engagement index (EEI)**. For the beginning, you can start with a simple survey.

Ideas for performance metrics are lead time, time from incident to recovery, number of releases, and test code coverage. To be pragmatic, use the DORA metrics for the beginning.

To measure activity, you can take the following:

- Deployment frequency
- Lines of code
- Commits
- Number of pull requests
- Number of reviews
- Issues/tasks completed
- Story points solved

Those numbers are not always comfortable to gather. In some countries, it would be explicitly forbidden to evaluate someone's performance and it is only allowed for a certain number of people. This also puts SPACE frequently into criticism. Therefore, if you would like to use SPACE, and we have to admit that the metrics are helpful to gather insights into the development cycle, our key customer, you should talk to your data protection team, who are responsible for setting this up right.

Similar issues occur with the communication and collaboration dimension. If you don't want to look into the user statistics of your communication tool, merge requests, merge times, or the number of comments at commits, merge requests, or issue tickets, are a good beginning. You can also evaluate the quality of meetings by asking for feedback.

The last dimension of efficiency and flow can be based again on DORA metrics. In addition, it is helpful to evaluate how many hand-offs you have in a development cycle, how many focus hours a developer really has, and what their velocities are.

The SPACE framework can be used at individual, team, and organization levels. The following diagram from Michael Kaufmann's book, *Accelerate DevOps with GitHub*, gives further examples of those levels.

	Satisfaction & Well-being	Performance	Activity	Communication & Collaboration	Efficiency & Flow
Individual	• Developer Satisfaction • Retention	• Code Review Velocity	• Focus Time • # Commits • # Issues / PBIs • Lines of Code	• Code Review Score (quality) • PR Merge Times	• Knowledge Sharing • X-Team Reviews
Team	• Developer Satisfaction • Retention	• Velocity (shipped) • Delivery Lead Time	• Cycle Time • Velocity (done) • # Issues / PBIs	• Code Review Engagement • PR Merge Times • Meeting Quality	• Code Review Stale Time • Handoffs
System	• Satisfaction with Engineering System	• Velocity • Lead Time • Customer Satisfaction • MTTR	• Deployment Frequency	• Knowledge Sharing • X-Team Reviews	• Lead Time • Velocity

Figure 2.18: Examples of SPACE metrics by Michael Kaufmann [16]

With the SPACE framework, we now get deep insights into the development, which is helpful for us as platform engineers, and it extends the DORA metrics. We can build those metrics on each other. This allows the product owner, architects, and platform engineers to evaluate if they do the right thing in the right way. However, we are missing input from the **developer experience** (**DevEx**) experts.

DevEx framework

SPACE and DORA lack the DevEx perspective. Also, while we are looking into satisfaction and flow, it doesn't give us a full picture of our platform engineering team. This is where the **DevEx framework** comes into play *[17]*. It limits the perspective back to just three dimensions. All of them together build the foundation for DevEx and can extend its scope of action.

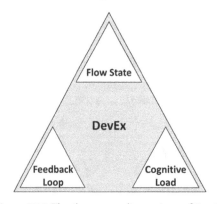

Figure 2.19: The three core dimensions of DevEx

From a developer's perspective, the flow state is when they focus 100% on writing code and literally forget about time, hunger, and other needs. The easier a flow state is to achieve and maintain, the better the DevEx.

We talked about cognitive load a lot. In short, a good DevEx includes a very low cognitive load.

Finally, the feedback loop measures how fast and with which quality a developer receives feedback on their action. The better this feedback is, the more continuous and actionable a developer can react with minimal friction and loss on the transaction.

To measure those is difficult as an answer lies between actual metrics and subjective feedback:

- **Flow state**:

  - **Metrics**: Re-opened commits, number of change requests, number of committed per pull request.

  - **Subjective feedback**: How often do they feel interrupted? Do they feel continuously stressed or under pressure? How high is their level of decision autonomy?

- **Cognitive load**:

  - **Metrics**: Number of reopened bugs, debugging time, number of dependencies for a component, and time to solve technical problems

  - **Subjective feedback**: The time spent coding versus the time spent on other tasks; the technical scope for their component, and how often they have to switch context

- **Feedback loop**:

  - **Metrics**: Cycle time, deployment frequency, **mean time to resolution** (**MTTR**), number of bugs, and level of test automation

  - **Subjective feedback**: Effectiveness of code reviews, internal team communication, and quality of feedback

As you can see in those examples, there is a lot of room for interpretation. When you define your platform's KPIs, you have to ensure that those metrics are clearly described. You have to be especially precise by explaining how you interpret which feedback and metrics in which way.

> **Important note**
> Architects must understand those KPIs and their meaning so that they can effectively improve the platform.

Using performance metrics

With the previous framework, some performance metrics were highlighted, such as the number of issues closed or story points. These metrics work well on a team level but can also lead to many misunderstandings.

However, there is a way that's not abstract and unclear, like a story point, and it is a bulletproof approach to talking with users and stakeholders about the platform and its efficiency: costs. I know this is not an innovative KPI, yet it is often done incorrectly by many teams and is feared by product owners. What I also don't mean by it is to report the team's costs, infrastructure, or licenses. You can use the cost metrics as proof of the performance of your platform and the platform team.

Cost per change

Cost per change is an efficient way to highlight a platform's improvement and power. Within the end-user community of the cloud-native ecosystem, we see examples of this KPI used to discuss how well a platform is running.

Let's compare two different values for the cost per change:

- **$500 per change**: $15,000 per month or $180,000 per year by 30 changes monthly
- **$80 per change**: $2,400 per month or $28,800 per year by 30 changes monthly

Now, obviously, there is a drastic difference between both numbers. But you should consider that you will always start with higher costs and very few changes. Ideally, the platform team and the platform become better over time to stabilize overall costs, but the number of changes increases. This leads to lower costs per change. The following chart represents this in order to visualize it better. The dotted line shows an example of the overall costs. They increase slightly over time as the platform becomes more effective. The dark line shows the changes per year, and the dashed line is the cost per change. As you can see, the cost per change drops drastically the more changes you are able to make. It is a very basic number, but a strong communication tool for your platform and its platform engineering team.

Figure 2.20: Total costs per year example

Costs per project/product

Related to cost per change is cost per project, product, or application. We can look at these costs from two different directions:

- **Total shared costs**: Looking only at aspects such as team costs, additional efforts, or costs that are not causation-based

- **Balanced causation-based costs**: The causation-based costs include, besides the shared costs, the real costs of infrastructure and services consumed by a project

To highlight the performance of the platform and the team, I would recommend going with the total shared costs. Such numbers are simple to get and split between the projects so that you can track them on a monthly basis without too much effort. An alternative to that is the *introduction costs per project*. For this metric, you need to calculate the effort that goes into the onboarding for each project using the platform. However, I personally wouldn't spend the effort on these metrics. They are more rhetorical numbers and don't provide insights into your platform performance, but more about how efficient you are in the onboarding.

User/product overhead costs

The overhead costs perspective is very similar to the cost per change. Here, you sum up all direct and shared costs and reduce them by only the consumed resources. So, it is important to differentiate between tools such as agents, network traffic, and maybe personal user-related costs from the actual resources used by the product. Calculating this is complicated, but the result is a performance metric of your platform's and platform team's overhead costs.

It's a metric that splits the minds. On the one hand, it should be as small as possible to be able to sell the platform internally. On the other hand, that number represents the costs of the actual value of the platform. So, I believe we would need to hide the middle ground that is not too costly but also not too expensive.

Summary

This chapter taught you about the defining platform principles and how to develop your platform team purpose. You now have the tools in your hands to document and design a platform architecture, followed by end-user examples for different platforms. This chapter gave you additional perspectives on the capabilities of platforms. With the approach of the TVP, you learned how to focus on single milestones without overarching and to adapt faster to the demands. Finally, we introduced you to a couple of approaches to measure your platform adoption and performance metrics.

In *Chapter 3*, we will discuss the details of building the foundation for platforms. First, we will introduce you to our reference and example company and explain certain details. You will learn about infrastructure foundations, multi-cloud and **Software as a Service** (**SaaS**) challenges, and creating a reference architecture for the foundation.

Further readings

- [1] *Cost of Delay: the Economic Impact of a Delay in Project Delivery*: https://businessmap. io/lean-management/value-waste/cost-of-delay

- [2] *Evan Bottcher, What I Talk About When I Talk About Platforms*: https://martinfowler. com/articles/talk-about-platforms.html

- [3] *Miro Template, The Platform Engineering Purpose Canvas*: https://miro.com/ miroverse/platform-engineering-purpose-canvas-template

- [4] **Team Topologies**: https://teamtopologies.com/

- [5] *PlatformCon2023, Platform as Code: Simplifying developer platform design with reference architectures*: https://youtu.be/AimSwK8Mw-U?si=1CDAJb1gLtlCgeyH

- [6] **PlatformEngineering.ORG toolings**: https://platformengineering.org/ platform-tooling

- [7] **CNCF landscape**: https://landscape.cncf.io/

- [8] **Wasm examples sources**:

    - **General page**: https://webassembly.org/

    - **A simple Kubernetes operator to run Wasm – kwasm**: https://kwasm.sh/

    - **An enterprise-grade operator and development toolkit for Wasm**: https://www. spinkube.dev/

- [9] **ThoughtWorks platform tech strategy layers**: https://www.thoughtworks.com/insights/blog/platform-tech-strategy-three-layers

- [10] **The Lean Startup**: https://theleanstartup.com

- [11] *ICE Scoring Model*: https://www.productplan.com/glossary/ice-scoring-model/

- [12] **Startup lessons Google Maps**: https://www.startuplessonslearned.com/2010/09/good-enough-never-is-or-is-it.html

- [13] **GitLab DORA metrics**: https://docs.gitlab.com/ee/user/analytics/dora_metrics.html

- [14] **DORA report 2023**: https://cloud.google.com/devops/state-of-devops

- [15] **DORA tutorial**: https://keptn.sh/stable/docs/guides/dora/

- [16] *Michael Kaufmann, Accelerate DevOps with GitHub*: https://www.packtpub.com/product/accelerate-devops-with-github

- [17] **DevEx framework**: https://queue.acm.org/detail.cfm?id=3595878

3

Building the Foundation for Supporting Platform Capabilities

Solving a problem users have, designing for good user and developer experience, and avoiding technical complexity are the foundational steps toward successful products and successful platform engineering.

In reality, many projects fail because those foundational principles are neglected: we see architects getting lost in technical details and losing sight of the problem they need to solve. Too often projects fail because the end user wasn't continuously involved throughout the creation process of a new product. Frequently, architectural decisions are made without considering how this new product will fit into the existing ecosystem, processes, and skill set of the organization. All of this leads to a shaky foundation with limited potential for success.

In this chapter, we'll walk through the mandatory steps and processes of defining a solid foundation of a platform that can grow from an initial set of features to key enterprise-supporting platform capabilities.

As such, we'll cover the following main topics in this chapter:

- Financial One ACME – our fictitious company
- Overcoming platform complexity by finding the right perspective
- Considering existing processes and integrating a new implementation
- Designing the infrastructure architecture
- Exploring multi-cloud, multi-SaaS, and the fragmentation of capabilities
- Exploring a reference architecture for our platform

Financial One ACME – our fictitious company

In this section, we'll learn how to understand the requirements of users in an engineering organization, how to balance the requirements of different teams, and how to decide what capabilities should be included in a platform and which ones not.

To make this more applicable and real, we're going to introduce to you *Financial One ACME*. While it's a fictitious company, the history, technology challenges, and teams we'll introduce and use throughout the remainder of this book are what we've seen in many organizations that we've worked with in the past years. Having such an organization allows us to better explain how to apply the theory we'll present in this book to practical actions.

Let's have a look at some important details about *Financial One ACME*:

- **A brief history of Financial One ACME**: Financial One ACME has been a leader in providing software solutions for the financial services market. The company started in the early 2000s with a classical 3-tier application architecture (Windows rich client, application server, and database) that customers installed in their data centers, with new releases of their software shipping twice a year.

  Over the years, the Windows rich client was slowly replaced with a web client.

- **Providing a self-hosted SaaS and on-premises offering**: In 2015, the demand for a SaaS-hosted version increased. Instead of re-architecting the product to support multi-tenancy for a SaaS-based solution, the decision was made to simply host the Application Server and Database for each SaaS Customer separately in the existing Financial One ACME data center. This solved concerns about data residency and data access. However, it meant that with a growing number of SaaS-based customers, the operational overhead grew as each customer (=tenant) had a production deployment on their separate VMs.

  Additionally, IT operations had the challenge of continuously expanding data center capacity for production environments to keep up with business growth. Due to different customer sizes, they also had to do individual capacity planning for each tenant to properly size and not oversize environments.

  Development teams, on the other hand, owned the pre-production environments. This included everything from the development workstations, build servers, and pre-production testing and staging environments.

- **Moving to monthly releases**: From 2015 to 2020, the release cadency was constantly improved, leading to a monthly release cycle. Those releases were built and tested by the development teams and then provided to IT operations to deploy into SaaS, as well as to those customers that still ran the software on-premises. This change in speed resulted in a situation where not every customer wanted to update their software that frequently as it had to be aligned with their internal change request processes. Some customers were up to six releases behind the current build, which put an extra strain on the development team to support all those older versions.

- **The lift and shift to the cloud**: In 2020, the business expanded into new regions, and with that, the requirement grew to offer the SaaS offering in new regions. Instead of building more data centers, the company decided to *lift and shift* their existing production architecture into the public cloud vendors' compute offerings.

  This move also included new processes and guidelines on how to update software or access data on those cloud servers – for example, how to access logs from server X that belong to Tenant Y!

- **Outperformed by new FinTech companies**: During the pandemic, new FinTech software companies emerged that were *born and architected for the cloud*! This put pressure on Financial One ACME as those new SaaS-only companies didn't have to maintain a legacy architecture to support both SaaS and on-premises. Their architectures were also multi-tenant and multi-cloud, which allowed them to run more cost-efficiently than the Financial One ACME counterpart offering.

Leadership made a strategic decision to re-architect their offering while also improving the efficiency of developing, supporting, and operating the existing platform until every customer could be migrated over to the future SaaS-only offering.

And here we are: it's 2024, so it's time to think about how to help this organization modernize and re-invent themselves. Kubernetes and cloud native are the future tech stack! The hope lies in platform engineering to improve the work of engineers that work on the future stack as well as maintain the old stack. This is where we come in – the newly formed platform engineering team!

Now, let's understand who the potential users of our future **internal developer platform** (**IDP**) might be:

- **Development teams**: Managing all pre-production tools and environments
- **IT ops**: Who has been managing on-premises data center and cloud compute resources
- **DevOps**: The team responsible for deploying and operating the existing SaaS-based production environments from an application perspective
- **Quality engineers**: The teams focusing on testing new software before it gets promoted to production
- **Site reliability engineers (SREs)**: Teams focusing on resiliency, availability, and help with reporting and enforcing SLAs and SLOs
- **Technical docs**: The team who's preparing end-user-facing documentation to accompany every release, along with release notes, new features, and how-to guides
- **Some others**: This includes **database admins**, **project managers**, **product owners**, **ProdSec**, and others

Now that we know who *we* as a team are and who *our* potential users are, let's look into *how* we need to build a platform that solves *their* pain points!

Overcoming platform complexity by finding the right perspective

"We spent months building our new platform. Devs hate it! Help me understand why!"

We don't want to end up in a situation where we feel like we have to post such a question on a public discussion forum. This heading is – believe it or not – from a real post. If you want to learn more about this story, then read the **Reddit** post *[1]* provided in the *Further reading* section.

So, how can this happen? The reason why many platform engineering initiatives fail is the same reason why other product development initiatives have failed: somebody had a great idea – built or made somebody build a product – but ended up finding nobody that saw the benefit of this new product as it didn't solve any real problem people have.

The mistake many make is to not validate the initial idea with their potential end users that they think would benefit from that solution. If you don't find a critical mass of users that have a problem that's solved by a new product, then it's better to not build a product at all as it's bound to fail from the start.

If you've done any product management in the past, you're probably thinking, *"But this is Product Management 101!"* Fully agreed! However, not every team that's tasked with building a new platform has previous experience in product management. Many teams that we've seen happen to find themselves in a situation where they can start building a platform without seeing the parallels to building any regular product. Many tasks go with building a product before even starting to build *the product*.

Applying basic product management – "Don't give your users a faster horse"

The right approach to successful platform engineering initiatives is aligned with what successful product teams have done in the past:

- *Identify* a problem that a large enough user base has
- *Understand* why this is a challenge and what the negative impact is right now
- *Research* on why this problem hasn't been solved already
- *How* you would quantify the benefit if you could solve this problem

To answer those questions and go a step beyond the basics, we suggest talking with your potential end users. Let them tell you about the actual problem they have and give them the freedom to explain a solution that gets them to the ideal way of getting their work done.

As you listen, make sure you aren't constraining yourself with any technical limitations or challenges you're aware of right now. There's a famous quote attributed to *Henry Ford* who allegedly said, "*If I had asked people what they wanted, they would have said faster horses.*" You'll learn that your users typically easily describe the problem they have. However, the solution they often come up with is constrained by what they think is possible or by the technical limitations they're currently aware of.

Now, we aren't going to start by coming up with a revolutionary new way for software engineering that takes months or years. However, this mindset is important as it's a step toward solving problems that others have failed to solve in a way that the solution is widely adopted!

To start, let's think about how to *solve this problem* with the *simplest and fastest* technical solution. *Don't over-engineer* from the start to come up with the best or most revolutionary technical implementation. While this should be an inspiring goal, our initial goal is to get *fast feedback and validation* on whether our proposed implementation solves the pain of our potential end users.

To get this fast feedback, you need to do the following:

- *Deliver* a solution quickly
- *Show* it to your potential end users
- *Get* continuous feedback
- *Refine* based on the feedback
- *Keep* iterating through these steps

This process continues until you can prove that your users would be willing to use your solution as it improves their way of getting things done.

Avoiding the "sunk cost fallacy"

Not every project will lead to success, no matter how hard and often you try and iterate. The **sunk cost fallacy** is a known problem pattern that we see in everyday life decisions, as well as when in software engineering. It highlights the problem that organizations are continuing to invest in a strategy because of the existing investment already made, even when it's obvious that it would be better to stop that investment as it has no chance of success. There's plenty of material on this that you can read up on such, as the **Sunk Cost** article *[2]*.

Therefore, it's important to define when to stop your iterations. If the moment of "*Users love the solution*" doesn't happen within a certain amount of time, you have to be ready to pull the plug and stop the initiative. To do this, you need to set yourself milestones on reaching that **validation point** of your solution. Remember, as mentioned previously, you don't want to end up with a post that says "*We spent months building this platform. Devs hate it! Help me understand why!*"

Steps to building the thing that users need – a real-life example

Let's take a step back to the beginning. What is it that we need to build? In product management, there are many different memes (take a look at *Figure 3.1* for one such example) about how the things that are being built differ from what the user needs. The same is true if you search for memes on overengineering. They all come up with the same conclusion: *Building something without first understanding what the user needs!*

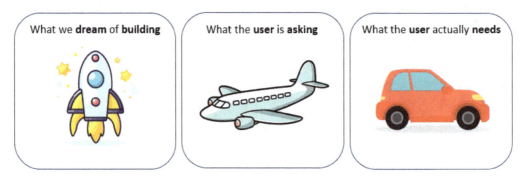

Figure 3.1: The overengineering paradox

It's possible to avoid a situation where we end up building something that doesn't solve the real problem or solves it in a way that's far too complex and there's no return on investment.

The hardest step in every journey is the first step. In our case, it's understanding *who* our potential platform users are and *what* the *real pain point* a platform can solve is!

To see this in action, let's walk through several steps on how we would approach this for Financial One ACME!

Step 1 – understand the real pain points of your users

In hallway conversations, the development teams often complain that analyzing problems in their software was so much easier in pre-production compared to production environments. They have full access to all the logs from their build server (Jenkins) and testing tools (Selenium and JMeter), as well as the environment where they deployed their software. They could easily increase log levels or quickly deploy a new version with more log output to triage problems faster.

When problems are detected in production, analyzing the problem is a completely different story! The development teams must ask IT ops for permission to get access to the logs by opening up a Jira ticket. That sometimes takes hours as the IT ops team is typically overwhelmed with many other tasks. The IT ops team also doesn't have the inside knowledge about where the software is writing all the logs to and the development teams don't always provide this information in the initial ticket either.

That's why it often takes several iterations to capture the desired logs and upload them to the IT-ops-owned tool to share production-relevant data with other teams. Changing log levels is also not that easy. Production changes like this have to follow a special change approval process. As its a software change, it's handled by the DevOps team, which slows things down even further!

For the development teams, this means that instead of just remoting into a server to access and analyze the logs on the spot, those logs have to be analyzed through the IT ops central production data storage tool. However, the development team isn't super familiar with this as they don't work with it regularly. That centralized tool also has its own permission system that was initially set up to prevent unauthorized access to restricted data. As that system isn't integrated with the same authentication and access control system developers use, it's often out of sync with the current team assignment. This is leading to situations where individual engineers in a team can't access the logs they need, leading to additional cycles with IT ops and DevOps or using a non-approved shortcut by simply asking a dev colleague who happened to have access to the logs!

As you can see, there's lots of frustration and pain in the development team. But as you can also imagine, there's also lots of frustration on the IT ops and DevOps teams. Their job of operating production is constantly interrupted with tasks to look up and provide access to data from production or approve a log-level change. There's constantly a lot of back and forth to understand what data is requested, where to find that data, and who should have access to this data.

Lots of conversations typically go into understanding the full picture, understanding the pain points on both sides, and having enough details to start thinking about a better way of doing this. When you apply this in your organization, plan enough time and notify people you want to talk to ahead of time so that they can gather their thoughts to have those conversations!

To start proposing a solution, it's best to get an organized overview of the pain points of all sides, as shown in the following table:

Problem: Devs don't have direct access to logs in production		
Pain Points: Development team	**Pain Points**: IT ops	**Pain Points**: DevOps
Need to create Jira tickets to request access to logs in production.	Need to deal with Jira tickets requesting access to logs.	
Lots of time is spent on tickets until the IT ops team identifies the right log to capture.	Devs don't always provide enough information in tickets. Additional iterations are required to get all the necessary details.	
It's hard to change log levels in production during triage. Yet more change request tickets need to be created.		Working on unplanned change request tickets to increase log levels.

Inefficient log analytics. Devs are more used to the tool used in pre-production. The production tool isn't as intuitive to them and slows them down.		
Dealing with permission issues in the production log analytics tool.	Additional overhead to explain that ownership and permission information is not in sync with dev systems.	

Table 3.1: Organizing the problem and its pain points into an easy-to-consume table

Now that we have a list of pain points from both sides, it's time to think about a solution to those pain points, how much of an impact this solution would have, how much it would cost, and what the **return on investment (ROI)** would be!

Step 2 – quantify the benefit of solving those pain points

We start by quantifying the impact that a solution would have. This is necessary to justify how much time and effort goes into building a platform that solves those pain points. While the pain explained previously is real, we need to understand whether these are single occurrences or regular occurrences. The question we need to answer is, is it justified to invest weeks into building a platform that solves this problem?

Going back to the same teams, it's time to quantify the costs of those listed pain points in terms of time spent or actual $$. We can get those numbers either through educated guesses or numbers from their current time tracking (the best option). As the team is currently using tickets, we should be able to get the total amount of time spent on those tickets on both sides.

Here's a revised table with the additional cost impact!

Pain/Time Spent per Month	Development Teams	IT Ops	DevOps
Slow process to request log access	2 days	4 days	
Extra process to change the log level	0.5 days		0.5 days
Inefficient log analytics	2 days		
Workaround for permission issues	0.5 days		
Total Sum	5 days a month or 60 days a year	4 days a month or 48 days a year	0.5 days a month or 6 days a year

Table 3.2: Quantifying the benefit of each pain point and providing an easy overview

Now, this is a great overview with interesting stats to make our decision easier. If we can solve all those pain points for the problem that developers have, which is that they currently don't have easy access to the necessary log files in production, we can save up to 114 engineering days per year. That's a good starting point and a great argument to invest in improving the efficiency of our teams by investing in platform engineering!

Step 3 – propose a solution that improves developer experience

Now that we know that we can potentially save up to 114 **full-time equivalent** (FTE) engineering days per year, we should go ahead and come up with a proposed solution that we can present back to the users involved. We must not present the technical details of the solution but how the developers will experience the user journey. **Developer experience** is the key word here that gets talked a lot about in platform engineering. So, let's come up with a solution that gives our developers a new experience that will make them want to use our solution!

Like in product engineering, we need to involve our end users by asking for input on how they would like to interact with a future solution. Developers often favor an approach where they can do everything through code or a simple-to-use **command-line interface** (CLI). The important piece here is that we want to come up with a solution that fits into the current work process and tools so that our users don't have to learn yet another tool or change their way of working.

The proposal in our scenario leverages the *Configuration as Code* approach. Developers can specify log levels, log output, ownership, and notification channels in code. This could be a standalone YAML or JSON file or could be part of a Kubernetes deployment definition. The developers simply need to check that file into their Git repository. DevOps and IT ops can validate and approve the **pull request** (**PR**) to make sure that all the data is accurate. If a new alert comes in, or if someone requests logs on demand, the new platform engineering capability will get the right log files for that component that has a problem. Then, it uses the ownership and notification information to contact the development team with a link or a digest of the logs that are relevant to that situation. The following diagram visualizes the proposed end-to-end workflow and shows how it improves the experience for developers, IT ops, and DevOps:

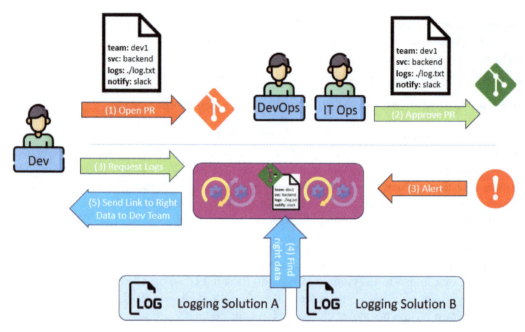

Figure 3.2: Improving experience through Configuration as Code for dev, DevOps, and IT ops

The proposed solution addresses all the current pain points while ensuring that only teams that own a component get to see their data. This solution is also extensible to make it work the same in production and pre-production environments. This would be a future iteration of this capability!

Step 4 – your first prototype

If the proposed solution is accepted, it's time to work on a prototype implementation. Prototypes are a great way to get fast feedback on an implementation that must not be technically perfect – yet! For our intents and purposes, a prototype is the best way as we want to validate that the problem users want to get solved can be solved in a way that they would start using our solution.

The first part of the prototype should focus on the *interface* to this new capability we are building. In our case, this is the config file we discussed earlier. A key consideration for the prototype is deciding whether the solution should solve the problem on the existing technology stack (3-tier app on cloud VMs) or whether this is a problem that we only want to solve as Financial One ACME moves their new cloud-native implementation. The goal should always be to provide the same developer experience, regardless of the underlying technology stack. However, when it comes to the implementation, the gain and effort might be very different.

To proof the prototype for both technology stacks, see the following Configuration as Code files:

service-metadata.yaml	service-abc.yaml (Kubernetes Deployment)
``config_version: 0.1
environment: prod-useast-01
service:
name: fund-transfer-service
group: backend-services
version: 2.34
ownership:
team: dev-team-backend
notification: slack
channel: backend-log-feedback
observability:
logs:
level: info
location: ./logs/service-log.json```	```apiVersion: apps/v1
kind: Deployment
metadata:
name: fund-transfer-service
namespace: prod-useast-01
spec:
...
template:
metadata:
annotations:
owner.team: dev-team-backend
owner.notification: slack
owner.channel: backend-log-feedback
app.kubernetes.io/name: fund-transfer-service
app.kubernetes.io/part-of: backend-services
app.kubernetes.io/version: 2.34
observability.logs.level: info
observability.logs.location: stdout
...
spec:
containers:
- name: fund-transfer
image: "financeoneacme/fund-transfer:2.34"
env:
- name: LOG_LEVEL
valueFrom:
fieldRef:
fieldPath: "observability.logs.level"
...```

Table 3.3: Two declarative ways to implement the proposed solution

Both options allow the development teams to provide all the relevant information and allow DevOps and IT ops to validate it:

- **Service details**: Name, version, to which component it belongs
- **Ownership**: Team identifier, notification tool of choice, and channel of choice
- **Logs**: What the log level is and where logs are written to

In this Kubernetes example, you can also see how we can pass some of this information in already standardized annotations such as `app.kubernetes.io/name`, `part-of`, and `version`. We can also see that information such as log level can be passed to the container as changing the log level in this configuration file can also change the logs that are getting produced by the deployed container!

Getting early feedback on this proposed configuration file format can happen before any actual implementation is done. It gives development, DevOps, and IT ops the chance to come up with additional metadata they want, such as the *priority* of the service. This could be used to define the escalation process when problem alerts occur in production, or a fallback email address to send email updates to the team.

The heart of our proposed solution is not the configuration file but the automation process that can react to requests from developers but can also be triggered from an alert. The easiest solution would be a simple service that exposes a REST endpoint to be triggered for both use cases. That service needs to run in an environment where it has access to the following:

- The Git repository or the K8s cluster that holds "the configuration"
- API access to the existing log solution in the target environment
- API access to the notification tools

To start with a minimum viable prototype, we can define that we only support K8s, only support the production log solution, and only support a single notification tool, such as Slack. This allows us to prove the value through the prototype with the option to extend it to other config data sources, additional log environments, and a longer list of notification tools.

What's remaining is the definition of the REST API. Here, we should consult with the development and IT ops teams as they are the main users of that API. Developers use it to request access to logs on demand, while IT ops use it to call the API when a problem gets identified. The following table shows a sample REST API definition for both use cases:

Request Data on Demand	Notify about a Problem
`GET https://logservice/request?service=fund-transfer-service&environment=prod-useast-01`	`GET https://logservice/request?service=fund-transfer-service&environment=prod-useast-01&incident=PROD-1234`

Table 3.4: Examples of a REST API for the newly proposed service

While both implementations will try to find the right log files for those services and send them to the correct team, the Notify API also receives an incident reference, which allows our implementation to include that information in the message that's sent to the development team.

There are many more things we could do as part of the prototype of this solution:

- Determine the current SDLC efficiency measured and learn how to measure the positive impact we intend to have

- Providing a CLI, chatbot, or web UI

- Create a template Git repository for new services to include the new configuration

- Create audit logs to track who is requesting data

- Expose metrics to track the usage and performance of the API

- Implement proper authentication to call the REST API

- Implement rate limits to avoid any issues when calling the backend Git, K8s API, or Logging Platform API

The list could go on.

It isn't necessary to implement all of these at the start to validate the prototype and prove the value of such capabilities.

Apply basic product management skills to your platform project

What we've learned so far is how to understand the real pain of our users, how to quantify the benefit of building a solution for those pain points, how to propose a solution, and how to keep the users in a close feedback loop by building a prototype!

Now that we understand that we need a product mindset when building a new platform, it's time to expand our requirements gathering. Looking beyond the needs of the developers by understanding how the platform will fit into the existing end-to-end processes and tools will set us up for future adoption and growth in terms of capabilities!

Considering existing processes and integrating a new implementation

We've just talked about how to identify the real pain points of your users and how to pick good candidates for your first prototypes to get early feedback. However, requirements must not just come from your end users. We must look beyond simply providing self-service through a chatbot, template repository, or a new CLI.

We need to look at and analyze the whole value creation journey and where the new platform capabilities fit in. We need to make sure that we understand the current **software development life cycle** (**SDLC**) – the process the organization follows as new software gets developed, released, operated, and retired. There are a couple of questions we need to be able to answer:

- What are the criteria and process to introduce change in the existing SDLC?
- How is the current SDLC efficiency measured and how can we measure the positive impact we intend to have?
- Are there any regulatory requirements for new tools we need to adhere to?
- Who within the organization needs to be informed or needs to approve new tools?
- Do new tools integrate with existing systems for authentication, access control, auditing, observability, security, and resiliency?

The first step is to understand the existing process, how we can prove the positive impact we want to achieve, the requirements to extend it, and who the key decision-makers are.

Understanding the existing SDLC – "the life cycle of an artifact"

As platform engineering aims to provide platform capabilities that improve and change the way developers do certain tasks alongside the SDLC, it's important to understand the current SDLC in the organization. Especially in large enterprises, it's very likely that there isn't just one process but many that have evolved over the years. It's also very likely that not many – if any – know the current SDLC from the first requirement and creation of the artifact until the new software is operated in production until it gets replaced or retired. It's very important to not make the mistake of assuming we – or a single person – know the existing end-to-end process. Even engineers who have been working for companies for many years often live in their own bubble and only have a limited understanding of what typically happens from the inception of a new idea until that code gets shipped, operated, and eventually retired.

The artifact life cycle experiment – from idea, to git commit, to prod!

A simple approach to learning the end-to-end process is doing a little experiment.

As a kid, I'm you were as fascinated by rivers (small or large) as I was. I'm sure you've dropped a piece of wood or a small branch of a tree into the water and then observed it as the water carried it away. You probably ran along the river to watch that branch making its way to its final life cycle step: the ocean! Because that's where all water streams eventually end up. As kids, we didn't have the chance to follow that branch to the ocean. However, as engineers, we do have the chance to follow the full life cycle of an artifact: from the inception of the idea (a ticket in a requirements engineering tool) to its first git commit by a developer until the artifact gets deployed to, updated, or replaced in production.

We have two options to understand that artifact life cycle. First, we can pick an existing service or feature and do some forensics by analyzing all tickets, git commits, pipeline runs, test reports, emails, change requests, and incident reports, and with that learn the processes, tools, and people involved!

Another approach is to ask one of the development teams to create a "demo" or "non-impacting" feature. With the popularity of feature flagging, this could be a simple feature behind a flag that changes a small runtime aspect of the artifact. The benefit of this is that it imposes no risk in production but it allows us to learn everything there is to learn about the current SDLC and with that derive the current artifact life cycle in that organization!

Insights into the artifact life cycle

I have run several of those "*Let's understand the life cycle of an artifact*" workshops with different organizations around the world. The results were lots of insights and learnings about the people, process, and tools. This included the following:

- What *tasks* were involved from requirement to the first git commit to prod deployment
- *who* was involved – the different teams that were involved in developing, testing, validating, and promoting changes from development to production
- The *tools* that were being used along the way and whether there might have been different tools being used for the same task in different environments
- Which tasks along the process were *manual* versus which were already *automated*
- *Dependencies* on other tasks or teams to keep the change moving toward production
- The *time* it took for each task, the overall time, the wait time between tasks, and more

This insight allows you to also craft a visualization of the full SDLC or the full artifact life cycle. Such a visualization will become very handy once we enter discussions on how our platform capabilities will impact the current existing processes:

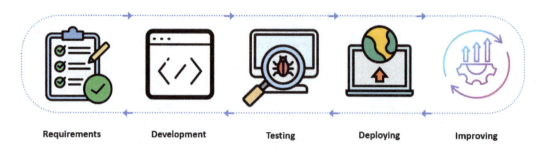

Figure 3.3: Understanding the life cycle of a software artifact

The artifact life cycle isn't limited to the initial delivery

While platform engineering initiatives often aim to improve the initial onboarding or delivery aspect of software components, we must not limit ourselves to this. This is why the term *artifact life cycle* is a good alternative to use as the life cycle doesn't stop with the initial development process. The life cycle of an artifact also includes operations, releasing updates, maintenance, or even replacing or retiring that artifact.

In our first example from Financial One ACME, we talked about an operational life cycle phase. This covered the tedious process for development teams to get access to log files in production to triage current problems. The following figure visualizes that life cycle phase and process!

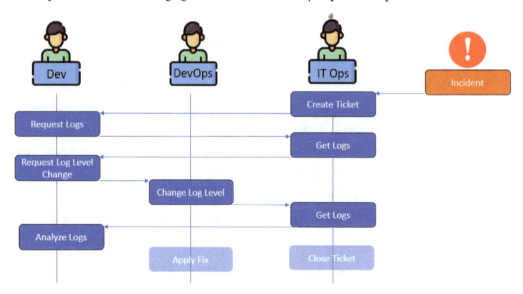

Figure 3.4: Life cycle of the incident response when accessing logs

Requirements from involved teams and existing processes

In our exercise to understand the flow of an artifact throughout the life cycle, we've learned a lot about the teams, existing tools, and processes involved. We've also learned about which tools our future platform capabilities may need to integrate, which teams we need to collaborate with, and – if we end up replacing or integrating with existing tools – what we need to do to not break anything that the current tool implementation provides.

Here are some examples of those findings:

- **Single sign-on (SSO)**: Every tool needs to integrate with the central SSO
- **Security**: Every tool needs to pass the software supply chain security guidelines
- **Auditing**: Every tool needs to create specific access logs
- **Service-level agreements (SLAs)**: Every critical tool needs to adhere to the company-wide defined SLAs on availability for critical services

So far, we've learned how important it is to understand the full process and life cycle of software artifacts. It's important to learn about which teams are involved and which existing processes we have to go through when introducing a new tool. We've also learned about additional future areas where a platform engineering capability can provide significant improvements to the software delivery and artifact life cycle!

Introducing life cycle events – measuring and improving the efficiency of the SDLC

We've already talked about the concept of the artifact life cycle. An artifact typically runs through a list of phases: requirement accepted, implementation started, pull request, artifact built, security scan finished, test complete, build validated, artifact promoted, deployment finished, feature released, problem detected, configuration changed, problem resolved, and artifact retired. While every organization will have slightly different life cycle stages or phases, we can visualize this nicely using the DevOps infinity loop. To trace the artifact along its life cycle, we encourage you to log every step along the way as a life cycle event. The following figure shows an example of those events and some of the metadata the involved tools and teams should add to better understand the full flow of an artifact, from its initial requirements to operations:

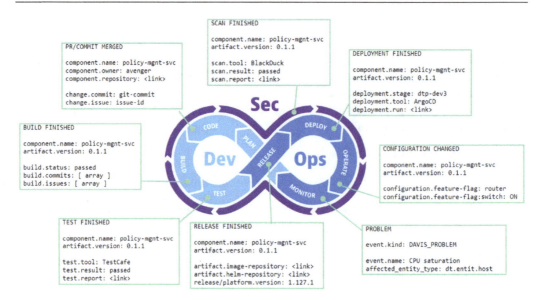

Figure 3.5: The SDLC and its artifact life cycle events

The idea of life cycle events isn't new. Many organizations are implementing the concept through different means, such as by adding log output into their CI/CD pipelines at the beginning and end of the pipeline, including metadata such as the timestamp, Git repository, pipeline, created artifact, and success status. The Continuous Delivery Foundation has been working on the **CDEvents** *[3]* spec, which defines a vocabulary of events that allow tools to communicate in an interoperable way. CDEvents is a great starting point that can be extended to cover the full life cycle of artifacts, as proposed previously.

There are many benefits of standardizing events like this:

- **Every artifact can be traced through its life cycle**: This means it can answer questions such as who was involved in the creation of an artifact!

- **Management of releases**: What version of a release artifact has been deployed and *when* by *whom* in *which* environment?

- **The life cycle stages can be measured, which gives us DORA metrics**: How long does it take from initial requirement until first deployment (lead time), how many deployments do we have (deployment frequency), how many deployments result in a problem in production (deployment failure rate), and how long does it take to fix a problem in production (time to restore service)?

- **Compliance**: It allows us to identify whether some artifacts have skipped important steps such as security scans, resiliency testing, and more.

- **Interoperability**: This specifies whether we use Jenkins or GitHub Actions to build artifacts. If all tools generate the same type of events, we stay in control of the life cycle.

Before you go off and define your life cycle events, have a look at the existing initiatives in the open source communities, as well as what the vendors in the delivery, observability, and **application life cycle management** (**ALM**) space are doing. There's no need to reinvent the wheel as standards are currently in the making!

Presenting the value proposition for improving the existing SDLC/DORA

Understanding the existing SDLC and making the life cycle of an artifact visible will be invaluable insights for our ongoing platform engineering initiatives. This will also be eye-opening for all members of the engineering organization as many of them most likely never got that overview to see where the current process provides opportunities to be improved.

The gained insights allow us to work on a value proposition that will improve the day-to-day life of not just development teams but ideally many other teams along the SDLC. This is why understanding and measuring the current process is so important – it allows us to not only propose a new solution but also present the improvements backed by hard facts, such as improving lead time, deployment frequency, deployment failure rate, or time to restore service (our beloved DORA metrics).

Regarding everything we've learned so far, how would this look in our log access use case for Financial One ACME? Our initial use case only focused on improving the access to logs in case of an error in production. Upon understanding the full end-to-end process, all teams involved, as well as existing processes, tools, and organizational requirements, we could propose the following solution and value proposition:

Proposal: Automating observability as a non-functional software requirement
Value proposition: Improve DORA metrics, reduce cognitive load, and scale best practices
KPIs and impacted teams: *Improve time to restore service by 50%* by automatically routing relevant observability data from production to the development team. This eliminates manual log capturing and forwarding for DevOps and IT ops.   *Improve lead time by 50%* by automating the process of capturing and analyzing observability data (logs, metrics, traces, and events) in earlier stages of the life cycle (dev, testing). The analysis will improve and automate the validation of new build artifacts and therefore reduce manual effort for quality engineers.   *Reduce production deployment failure rate by 50%* by automatically capturing problems in earlier environments by automating the detection of common problems (new critical logs, exceptions, slowness in the app, and so on) based on observability data.

High-level user journeys and collaboration:

As a *development team*, I commit an *Observability as Code* configuration file that includes information about the log source (for automated capturing), ownership (for automated routing), and important log patterns (to be used in automated validation and automated alerting).

As a *quality engineering team*, I collaborate and extend the log patterns to auto-detect regressions as part of test result analysis.

As a *DevOps team*, I collaborate and extend log pattern detection rules based on experiences and patterns seen in other development teams.

As an *IT ops team*, I collaborate and validate correct log sources and extend production problem detection based on relevant log patterns defined by development and DevOps.

Table 3.5: Formulating a proposal for our solution in a presentable layout

Have a clear proposal and value proposition
The chance of success will be higher if we can clearly articulate the benefits of our platform.

Now that we've learned how to craft our proposals so that they include everyone involved in the software delivery and artifact life cycle, it's time to think about how we should design the proposed solution.

Designing the infrastructure architecture

At this point, we've pitched our proposal. Once we get the initial buy-in from all teams and executive sponsors involved, it's time to move to the next stage: we need to think about designing the solution and how it fits with the underlying infrastructure to achieve all those non-functional requirements of our organization: resiliency, availability, auditability, security, mandatory integrations, and so on.

While we must not over-engineer from the start, it's important to be aware of all the requirements that will have an impact on our infrastructure decisions. Here are a couple of questions we need to be able to answer:

- How, who, and where are we deploying, updating, and operating the platform?
- Are there organizational requirements to run on certain infrastructure?
- Is there a requirement to run our solution across multiple geographical regions or even across multiple infrastructure providers?
- Are there infrastructure requirements to access and connect with other systems?
- How many users must our solution be able to sustain?

- What are our SLAs for infrastructure as well as for end users?

- How would we scale in and out?

In *Chapter 4*, we'll spend more time diving into architecting the platform core while using Kubernetes as the unified orchestration layer. Whether you end up deciding on Kubernetes as the underlying abstraction layer or using something else, you will have to answer these questions as they will impact some of the decisions you will make. So, let's dive into getting some answers on how that would impact our architectural decisions!

Avoid the ivory tower approach – we own the platform!

Before answering the infrastructure-impacting questions, let's answer the fundamental ownership question:

> **We own the platform!**
> The platform engineering team owns the platform as a product end-to-end!

As our platform will provide capabilities to make delivering and operating software components easier, our top goal must be to follow our own best practices and ideally deploy the platform with the same golden paths and self-service capabilities we want our end users to use. Some organizations call this *being customer zero* or *drink your own champagne*. If we don't follow that mantra and just build something, throw it over the wall, and hope somebody else operates it, we miss the whole point of having a modern product mindset for platform engineering.

We need to feel the same pain of software delivery and operations that our users currently feel. This will be even more motivation for us to build platform capabilities that make our lives as software teams easier.

We need to avoid an *ivory tower approach* where we're forcing best practices from the top down without following those best practices ourselves. If we do this, we may end up with a famous Reddit posting, similar to the one mentioned at the beginning of this chapter: "*We spent months building this platform. Devs hate it! Help me understand why!*"

Now, let's go back and answer some of those infrastructure questions that were raised earlier that may influence our architectural decisions.

Organizational constraints – existing infrastructure requirements?

We need to find out whether there are constraints on using existing infrastructure services. When it comes to enterprises, contracts with infrastructure or cloud providers will likely exist. If such constraints exist, it will play a role in our decisions – for example, do we need to run and operate our own K8s cluster on-premises or can we leverage a managed service from a vendor? If we're constrained to a certain cloud vendor, this also means we're potentially constrained to use their service offerings (storage, database, caches, and more).

Regarding that area, we also want to look into access control, ingress and egress, and cost constraints!

It's important to know about all such organizational constraints before finalizing the decisions on infrastructure and architecture!

Connectivity constraints – interoperability requirements?

Our platform capabilities will require us to connect to and interact with other existing systems. That ranges from having access to SSO, your Git repository, CI/CD pipelines, observability, orchestration layers, cloud APIs, and more.

Depending on where the platform will run – that is, on-premises versus the cloud – this will have an impact on how the platform can connect with all those tools or how those tools can connect back to the platform.

We need to think about firewalls, pull versus push connectivity, and API rate limits and costs as we need to ensure resilient interoperability between all those systems.

It's important to know all those connectivity constraints before finalizing the decisions on the infrastructure and location where our platform resides regarding all other systems we need to connect to.

Resiliency constraints – SLAs and other non-functional requirements?

The goal of our platform is to improve day-to-day work for a variety of internal users (developers, DevOps, IT ops, quality engineers, app support, and more). This means that our platform needs to be available and working every time our users need it. In global enterprises, this could mean operational resiliency and high availability 24/7. If the platform isn't available, our engineers can't do their critical work, such as releasing a new version of a piece of software, patching a security issue, or scaling their workloads to handle increased end user demand!

So, we need to understand the non-functional requirements on our platform, such as the following:

- Availability – for example, 9 A.M. to 5 P.M. in a single time zone or Monday to Friday across all time zones
- User experience – for example, the acceptable end user performance of a system must be guaranteed, with up to 100 concurrent engineers using the platform capabilities

Those non-functional requirements also mean we need to think about how we scale up and down, as well as scale out and in. We need to define whether we provide our platform centrally so that it serves all geographies around the globe or whether we need to deploy the platform components across the different regions to better fulfill the user experience requirements on acceptable performance.

Dynamic, horizontal, or vertical scaling is a topic that we'll look at in more detail later in this book when we dive into architecting the platform's core with Kubernetes as a unified layer.

Now that we have answers to all those questions, we'll be able to make better-informed decisions about infrastructure and architectural choices for our platform and its capabilities!

Exploring multi-cloud, multi-SaaS, and the fragmentation of capabilities

When we look into the fragmentation of capabilities and how that interoperates with the platform as a service and its scalability, one thing comes to our mind as a critical function that may not be the most obvious at first: the *IDP should be multi-tenant, not just multi-user*.

Multi-tenancy and ownership as a capability of our platform

Multi-tenancy is usually thought of in the context of a production application. For example, our platform client Financial One ACME is looking to turn their single-tenant product into a multi-tenant product. This will help them achieve higher profit margins, give them less operational overhead, and a slew of other business wins. The same data separation that exists within a production application in a highly regulated industry should exist at every level of that application life cycle, including the IDP. This is not only best practice but may be necessary to gain certain security and compliance certifications.

Since production data can make its way into tests, ensuring that the default access level of every user on the platform is as restrictive as possible by default helps to ensure that should data end up somewhere undesired, the surface level of exposure for that data is as minimal as possible. Additionally, the full isolation of multi-tenancy ensures that secrets that each team may need are isolated from each other, further mitigating any risk surfaces.

We'll explore the benefits of security and multi-tenancy in the platform further in *Chapter 7*. However, it's important to note that multi-tenancy can also help reduce the cognitive load of the platform's users. We'll cover how more fully in *Chapter 6, Building for Developers and their Self-Service*.

Multi-tenancy also means that any architectural decision we make – any tools we include in our platform – also support multi-tenancy. Take observability as an example – if we capture logs, metrics, traces, and events for the core platform, as well as the applications our users are deploying, we need to make sure that this observability data can be "separated" into its individual tenants. We've already discussed that our platform must include ownership as metadata when deploying our core platform, as well as when we provide templates for self-service onboarding of new apps. Let's consider an example deployment definition enriched with ownership metadata, as discussed in the *Overcoming platform complexity* section. That ownership metadata can be used to enforce access control to any data captured for that deployment, including using it for filters when querying for your metrics or traces:

service-abc.yaml (Kubernetes Deployment)	Using Ownership Metadata with Observability!

<pre>apiVersion: apps/v1
 	
kind: Deployment
 	
metadata:
 	
  name: fund-transfer-service
 	
  namespace: prod-useast-01
 	
spec:
 	
  ...
 	
  template:
 	
    metadata:
 	
      annotations:
 	
        owner.team: dev-team-
backend
 	
        app.kubernetes.io/name:
fund-transfer-service
 	
        app.kubernetes.io/part-
of: backend-services
 	
        app.kubernetes.io/version:
2.34
 	
    ...
 	
    spec:
 	
      containers:
 	
      - name: fund-transfer
 	
        image: "financeoneacme/
fund-transfer:2.34"
 	
    ...</pre>
	The annotations can be used as filters in PromQL queries, for example:
	
	<pre>sum(rate(container_cpu_usage_
	seconds_total{annotation_
	name="owner.team", annotation_
	value="dev-team-backend"}[5m]))</pre>

Table 3.6: How ownership metadata can be defined and used

This ownership information can be used to give access to observability data to those teams that own that component. It allows us to separate the data into its tenants while also allowing admins of our platform or owners of shared services to analyze data that goes beyond a single tenant. This is also an important capability as it's very likely we will run into cross-tenant issues. So, having all observability with the right context available will be critical to identifying issues and fixing them.

Data pipelines – whether using OpenTelemetry Collectors or commercial products – can also use this metadata to send the observability data to different backend storages to even separate the storage of that data based on tenancy!

The topic of ownership and observability will be discussed in more detail in the following chapters as it's a key enabler for many capabilities and attributes of our platform!

The considerations for running on multi-X

In the *Organizational constraints* section, we talked about understanding any existing constraints, such as whether we have the business or regulatory obligation to run multi-cloud, multi-SaaS, or – as some call it – **hybrid cloud**.

When Financial One ACME wants to sell into different global markets, there will likely be the requirement to operate our software out of a cloud vendor's most local region – for example, EU-WEST or APAC-SOUTHEAST. There might even be the requirement to run it across multiple regions or multiple different vendors to fulfill contractual obligations for high availability and resiliency while adhering to the standards that those companies need to obey.

For competitive or regulatory reasons, it's also possible that some cloud or SaaS vendors can't be used at all. Why is that? Well, if a potential customer of Financial One ACME considers one of those cloud vendors as a competitor or if one vendor doesn't meet some regulatory requirement, they wouldn't be able to purchase our software services.

This sounds like a lot of constraints leaving us with not much choice, doesn't it? Fortunately, there are solutions to solve that problem: in the coming chapters, we'll talk a lot about Kubernetes as the underlying abstraction layer. Building on top of this abstraction layer will make it easier – but not hassle-free – for us to move workloads to different regions and different cloud vendors.

As our platform and all of its components will not exclusively run on Kubernetes but also include other services, it's important to create a thorough checklist to avoid any architectural decisions that will be hard to change in the future. For example, if our platform has requirements for a document store service, we can either decide to run and operate a document database such as MongoDB on Kubernetes or pick an existing SaaS offering from one of the cloud vendors. Picking a managed service will reduce the effort on our end, but we need to make sure that each potential SaaS vendor that supports a document store service provides the same interface, similar performance characteristics, and similar cost structure. If we make the wrong architectural decision, it might be difficult to do the following:

- Change our implementation so that it supports a different API
- Have a performance and scalability impact on different vendors
- Increase the overall operational costs of our software

Centralized and decentralized platform capabilities

In the following chapters, we'll provide more examples of running our IDP and its components either as a central service, per environment, per region, or even per tenant (if that is a requirement). Your findings about organizational requirements will impact those decisions, such as whether our customers demand that they run in certain regions or aren't allowed to use a certain cloud vendor.

Centralizing components has the big advantage of them being easier to manage. Think about a central **continuous integration** (**CI**) system such as Jenkins or GitLab that runs pipelines triggered by pull requests to build new artifacts. A central instance of such a CI system is easier to operate, observe, secure, and manage. On the flip side, it means that all our users share the same system, which can easily lead to bottlenecks. The central system also needs to have access to all target environments, which may reside in different geographical regions or even other cloud providers. The number of dependencies to those cloud environments increases with every new team that onboards their application or with every new customer of Financial One ACME that demands our software to be deployed specially!

Decentralizing components have the advantage of being able to automatically encapsulate the data capture, storage, and access processes. These components are only used by those teams in those regions or environments, which limits the potential of patterns such as noisy neighbors, something we'll discuss in more detail in *Chapter 6*. The drawback of a decentralized approach is that it's harder to operate, observe, secure, and manage those components as the number of components will grow with every new environment. Automation can help here, but we can't downplay the additional level of complexity. Just think about whether a vulnerability is found in our CI system. Instead of patching it once in a central system, we need to patch every single instance across all our decentralized services!

Now that we've learned that multi-X – which includes multi-tenancy, cloud, and SaaS – will have an impact on our architectural decisions, it's important to find a good balance between the effort, benefit, and external requirements we have on our platform. All the knowledge we've gained so far allows us to come up with a good reference architecture for our platform, something we'll discuss in the last part of this chapter!

Exploring a reference architecture for our platform

Architectural diagrams are a great way to give a good overview of what a system provides to the end user and how the inner plumbing works. In this chapter, we discussed several important steps toward building the foundation for a future platform and its capabilities:

- **The purpose of our platform (solving end user pain)**: Who are our end users and what problem do we need to solve?

- **User interface/dev experience**: What is the ideal developer experience for our end users to best fit into their day-to-day work activity?

- **Core platform components**: Picking the right components to adhere to existing processes, tools, infrastructure, and constraints.

- **Our platform as a product**: Like any software product, our platform must be available, resilient (also work under heavy usage), and secure by default.

- **Success KPIs**: Use observability to measure and drive for adoption, efficiency, and productivity.

A different way to present this is through a high-level diagram of our platform. While the following diagram isn't complete, it can be a good reference for the platforms that you'll be building:

Figure 3.6: Reference architecture for our platform foundation (The image is intended as a visual reference; the textual information isn't essential.)

> **Our platform as a product has to have a clear purpose!**
> Start by visualizing and articulating the purpose before going into the technical details!

Let's walk through this diagram as you would when presenting it to your peers or end users. We'll start from the top down before we look into the aspects of observability, availability, resilience, and security.

The purpose – self-service for your end users

In *Chapter 2*, we gave an example of a platform principle that we called *self-service first*. We provided a detailed description stating, "*We will provide our customers with every platform capability as a self-service, focusing on their user experience and enabling self-determined software development.*"

This is exactly what's reflected at the top of the preceding diagram. Here, we listed all of the potential end users we learned about. For Financial One ACME, we have our development teams supporting them with use cases around easier onboarding of new applications or getting easier access to log files from the legacy systems in production. We also provide self-service for the DevOps, security, **site reliability engineering** (SRE), and application support teams by reducing the manual work and cognitive load through self-service capabilities.

There is only so much space on a high-level diagram like this. We could add examples for the different end user self-service features as text or – if needed – provide a more detailed version of this diagram for every end user and their self-service features!

User interface/dev experience

Every end user group is different. Even within end user groups we may have different skill sets that need to be considered when designing and implementing the user interface for our platform's self-service capabilities.

Development teams most likely prefer to stay in their preferred IDE, such as Visual Studio Code or IntelliJ. They're probably OK to edit YAML files to enable new self-service capabilities. Some may expect an IDE extension that will bootstrap the creation of those YAML files or an extension that provides code completion and schema validation to reduce the chances of making a mistake such as a typo.

On the other hand, other teams within the same organization might want a nice developer portal UI – something such as Backstage, which provides nice templating features for creating new components.

Then, you will have your automation engineers, who neither live in an IDE nor want to click through a wizard to create a new project. They want an API or a CLI they can use to automate tasks as this is the way they're used to working. They probably expect some Python libraries to call our self-service capabilities.

As discussed earlier in this chapter, every user has different skills and therefore also different expectations of what a good user experience looks like. Focusing on a good user experience is key as this will be a deciding factor on whether the platform will be adopted or not. This is why it's important to highlight the user interface and how it may look in our reference architecture as this is the way our users will interact with our platform!

Core platform components

While we all have our preferences for picking tools and technology, the right approach to picking your core platform components should not be by personal preference. It must be based on what self-service use cases (that is, features of our platform) we want to deliver through a user interface that delivers a good user and developer experience.

In the preceding diagram, we subcategorized the platform components into the following areas:

- **Delivery services**: Those are tools and services that focus on software delivery as this is going to be a primary use case of our platform. CNCF open source tools such as Backstage, ArgoCD, Flux, and OpenFeature, open source tools such as GitLab, Jenkins, and K6, or even commercial tools such as the tools from Atlassian would fall into this category.

- **Platform services**: As a key use case will be around delivering and operating software components, our platform should also provide core platform services that those software components need. That could include services such as caching or databases (Redis, MongoDB, or Postgres), messaging and eventing (Apache Kafka or RabbitMB), service mesh (Istio, Linkerd, or Nginx), and core services such as secrets management, container registries, policy agents, auto-scalers and more.

- **Platform**: While the platform doesn't need to run on Kubernetes, in the remainder of this book, we will focus on K8s as the underlying platform orchestration layer. However, Kubernetes must not be the answer to all platform engineering questions. The goal is to pick the right platform components to provide the self-service your end users require. This may lead you to run your platform on VMs or even pick a fully managed out-of-the-box SaaS solution!

- **X-as-Code**: Besides *self-service first*, we must also embrace an *Everything as Code* mantra. Whether it's defining deployments, observability, or ownership, as discussed earlier in this chapter, or any other software life cycle aspect, everything should be expressed in some type of code. This is where you want to pick tools such as Crossplane, Terraform, or Ansible. Crossplane in particular is emerging in the cloud-native space as a tool of choice when it comes to orchestrating applications and infrastructure in a declarative way. Also, make sure to validate your tool choices for delivery and platform services based on how *X-as-Code*-friendly they are, such as how they define **service-level objectives** (**SLOs**) or monitoring alerts with the selected observability tool!

- **Observability**: You wouldn't drive a car or fly a plane without any telemetry and your key indicators such as speed, altitude, or fuel gauge. We must also not operate any platform without observability built-in. Thanks to standards such as OpenTelemetry and projects such as Prometheus and Fluent, we have a lot of built-in observability signals in the components that we'll be using in our platform. Whether it's Kubernetes itself or tools such as Nginx, Reddis, Harbour, ArgoCD, or Backstage, all those tools emit metrics, logs, events, and/or traces that we can collect and send to our observability backend for automated analysis or alerting! While lots are already built in, not every tool we choose will likely provide all the data we need out of the box. Therefore, it's important to validate your tool choices based on their state of observability. For some tools, you may need to extract metrics from logs. For others, you may need an extension – for example, Jenkins provides an OpenTelemetry plugin to emit traces for each Jenkins Job execution. In other cases, you may need to rely on commercial observability offerings through agent-based instrumentation.

Observability will be covered more later in this book as it's the enabler for many disciplines, such as SRE, auto-scaling, incident response, and troubleshooting.

A platform that's available, resilient, and secure

A platform is a product! So, a product that we want to be used by our internal users has to be available when they need it, resilient when a lot of users need it at the same time, and secure to ensure our users trust the product.

To ensure that our platform is always there when our users need it, we need to apply the same architectural principles that we would apply to any type of software product that we want to become successful. For our platform, this means that all critical components of the platform also need to be available, resilient, and secure by default. Let's have a look at some examples and best practices for these three pillars.

Availability

Our users consider our platform available when they can use it for the self-service use cases we promised it would deliver. Remember our proposed *Request a log from production self-service* use case from *Figure 3.2*? If a development team follows that self-service use case, they expect that our platform will deliver the requested logs in a reasonable amount of time at any time, whether it's Tuesday at noon or Friday at 11 P.M.

That use case involves a lot of different tools, such as Git (where the config as code files are stored) and our homegrown solution, which offers a REST API to request specific logs. Both systems need to be available. Both systems also need to respond to requests in a specified amount of time with a valid response.

The best way to ensure availability is by following two steps:

1. **End-to-end use case monitoring**: As we would do for business-critical applications, we can set up synthetic tests to simulate the end-to-end user journey for that use case. In this case, we would validate that calls to the Git API work as expected by executing dummy commits and PRs and validating the response time and response code. We would also do the same for our REST API call to request the logs. However, this scenario also needs to wait until the requested logs are sent back to the development team's communication channel so that we can measure the full end-two-end time.

Both scenarios can be done through custom scripting, which we run as a cron job, or we can leverage existing synthetic testing solutions. You should validate whether your observability platform already provides synthetic tests as many of those vendors have this feature in their portfolio. This will also make it easier for our second step, which is monitoring and alerting on key quality indicators!

2. **Monitor and alert on key availability indicators**: As the platform team, we want to be notified if our platform self-service use cases are no longer working as expected before our end users start complaining. This is where monitoring and alerting on key indicators come in. Our synthetic tests already provide a good indicator, but we should set up automated alerting in case the Git API or our REST API starts to slow down or start responding to errors. On top of that, we can also alert on other leading indicators. Many tools in our platform – including Git, from our example – will produce logs as well as health indicator metrics. We want to be notified in case we see *ANY ERROR* logs on those systems. They can be early warning indicators. Most systems also have internal queues, such as request queues. We want to be notified if those queues keep growing since this is an indicator that too many incoming requests are piling up, which will eventually lead to performance and availability problems.

There are many other things we have to do to ensure systems stay available. We'll discuss those in the next section, where we'll talk about resiliency.

Resiliency

We just talked about the availability of our product. For our end users, the platform must always be available, whether they're the only user on the system or whether 1,000 others are trying to execute self-service tasks. They also expect availability, regardless of whether there are any issues in our internal infrastructure or our cloud services. This is when we talk about the resiliency of a system. It means that despite any unexpected events (high load, component failures, connectivity issues, and so on), the system itself stays available.

From an architectural perspective, there are many things we can do to make a system more resilient, thus leading to high availability. Let's consider our REST API example again. From a deployment perspective, we can do the following:

- **Load balance replicated deployments**: We can deploy our implementation using ReplicaSets, which leads to multiple actual Pod instances having to handle incoming requests. Requests can be load balanced with round-robin or other more sophisticated algorithms.

- **Implement geographical diversity**: To adhere to a global user base, we can deploy our implementation into regions where our users reside. This reduces the impact of network latency and distributes the load, depending on where our users are.

- **Provide automatic scaling**: Horizontal Pod Autoscaler (**HPA**) and **Kubernetes Event Driven Autoscaling (KEDA)** allow us to scale our implementation based on indicators such as CPU, memory, incoming requests, or even user experience. This allows us to avoid most resource constraint resilience issues.

- **Perform automated backup and recovery**: Not every issue can be averted. If disaster strikes, it's important to have automated backup and recovery options so that we can return to normal operation as fast as possible.

From a software architecture perspective, our REST API has several options:

- **API rate limiting**: We can limit the number of incoming requests overall or from a particular user or team. This avoids problems that may arise if someone accidentally (or on purpose) is flooding our system with too many requests.

- **Queuing incoming requests**: Event-driven architectures allow us to queue requests and work on them when workers are available. Workers can then also be scaled depending on queue length. While additional layers, such as request queues, will potentially impact request latency, this architectural pattern will increase stability, resiliency, and availability.

- **Retries and backoff to downstream systems**: As we also make calls to backend systems such as Git or the logging observability backend, we want to make sure we aren't impacted by problems on their end. For this, we can implement API retries (in case an API call takes too long or fails) in combination with backoff (increasing the time between API calls). Those strategies will increase our resiliency and also help our downstream systems.

Security

If our platform is available at any time, that's a great start. But like any software product, it has to also be secure. This is even more important for platforms as they enable our end users to create, deploy, and operate new software applications that also have to be secure. If our platform runs on a stack and tools that have known vulnerabilities or that can be hacked into, we're open to attackers leveraging this for software supply chain attacks or to get access to confidential information. Let's look at our REST API again. If we allow everyone to use that API to request logs from production, we may end up sending critical log information to a hacker.

That's why it's important to apply the strictest security guidelines to our platform, every component and tool we use, and every line of code we're developing. In *Chapter 7*, we'll cover the topic of building compliant and secure products in more detail, which is why we keep it at a high level here. We must consider our platform to be the most critical software in our company. Therefore, we must not deploy any components that have known vulnerabilities. We must implement proper access control in all our custom-developed APIs. We must use all the security scanning and alerting tools for our platform that we would also use for any business-critical software we run in production.

We're the platform engineering team, which means we're responsible for our platform's security, just as we are for its availability and resiliency!

Success KPIs and optimization

"We spent months building our new platform. Devs hate it! Help me understand why!"

With everything we've learned so far, we should be in a good position to not run into the problem we wanted to avoid from the start of this chapter. We want to build something that our users love to use as it makes their lives easier. But we can't just claim success based on a gut feeling or anecdotal evidence. We need to measure our impact and report our success based on KPIs and we need to leverage all the data we can observe from within our platform to keep optimizing those success KPIs, as well as the operational aspect of the platform.

We've already talked a lot about observability, both as a self-service feature of our platform and for our end users, as well as using observability to understand the inner workings of our platform to ensure availability and resiliency. Observing our platform components such as Backstage, ArgoCD, Jenkins, OpenFeature, Git, Harbor, Redis, Istio, and Kubernetes itself gives us a lot of insights into the success of our platform and its features. Let's dive into some key indicators, how to capture them, and what we can do with them to optimize our platform adoption and efficiency.

Active users

If tools such as Backstage are the developer portal of choice for our platform, we'll want to measure how many of our users log in to Backstage on a day or in a week. We want to measure how many new projects get created through the Backstage templating engine and how many new or updated Git repositories that results in.

For our Financial One ACME use case on log analytics, we can measure how often our REST API is called. As teams should identify with a team identifier or token, we can measure how many teams are using our feature and how often.

These are all good indicators of current adoption. If we see slow adoption, we can reach out with more education or utilize more one-on-one sessions to enable more teams. We can also use the existing adoption of individual teams and create internal success stories to promote the platform's capabilities to other teams.

However, it's important to start measuring how many active users we have on the platform and use this as a baseline so that we can set additional actions and grow adoption!

SLOs/DORA

A platform is a chance to promote best practices through self-service templates. This is where SLOs come in. Not only can we define SLOs for our platform components, such as availability, but we can take this as a chance to include SLO definitions in any new template for new software projects our end users are creating through our platform – we want their software to also be highly available, resilient, and secure.

On top of that, we can measure how many new software projects and releases are created and published with the help of our platform. We can look into the number of Jenkins job executions, how long they take, how often ArgoCD syncs, and how many of those deployments that end up in production meet their SLOs. These are all indicators that help us report parts of the DORA metrics back to the engineering teams. They need these metrics to show how efficient they are. This also helps us, as the platform team, to highlight how much more efficient teams are becoming with the help of our platform as we assume metrics such as *deployment frequency* or *lead time for change* will improve. We'll cover DORA in more detail in *Chapter 5*, where we'll dive into CI/CD automation!

Utilization/FinOps

Having a lot of users adopt our platform is a great goal and we've already learned how to measure it. Bringing more teams onto a centralized platform allows us to centrally enforce best practices around the right-sizing and right-scaling of deployments and optimizing the utilization of the underlying infrastructure, which results in optimizing costs for the platform and all the apps that get deployed through it.

In *Chapter 1*, we mentioned a new European regulation for reporting carbon emissions. This can also be done centrally as a platform capability by reporting actual resource utilization, the costs, and the calculated carbon impact of the platform, each platform service, and every application that gets deployed and operated through the platform.

There are many more such use cases that help us optimize the utilization of the underlying infrastructure to keep our costs under control. Overall, this means that our platform is a great place for our FinOps initiatives!

We hope you can apply a similar approach to creating an architectural reference architecture for your platform engineering projects. Start with a high-level overview that shows the purpose of the platform, gives an overview of the features and the user interface, gives insights into the core platform components, and indicates how you measure the success of your platform engineering initiative.

Summary

In this chapter, we learned how to approach platform engineering with a product mindset: find the real problem we need to solve, provide a simple quick solution to validate our implementation, identify how our platform fits into existing processes and organizational requirements, pitch a solution with the value proposition, which doesn't just focus on development teams alone, and then design for flexibility without going down the route of over-engineering!

We ended up with a high-level reference architecture that you can use to promote the purpose of the platform to your end users and any other stakeholder that needs to support this project.

In *Chapter 4*, we'll dive into the architectural details of a platform, the role of Kubernetes, how to integrate it with your existing services, and how to provide the platform's capabilities to your application and service development teams!

Further reading

- [1] Reddit Post: `https://www.reddit.com/r/devops/comments/stuep4/weve_spent_months_building_this_platform_devs/`
- [2] Sunk Cost: `https://developerexperience.io/articles/sunk-cost`
- [3] CDEvents: `https://github.com/cdevents/spec`

Join the CloudPro Newsletter with 44000+ Subscribers

Want to know what's happening in cloud computing, DevOps, IT administration, networking, and more? Scan the QR code to subscribe to **CloudPro**, our weekly newsletter for 44,000+ tech professionals who want to stay informed and ahead of the curve.

https://packt.link/cloudpro

Part 2 – Designing and Crafting Platforms

In *Part 2*, we will tackle the technical foundation of a platform and walk you through relevant decision points during the design process. To do so, you will learn about the four primary parts of a platform – the core components and infrastructure represented by Kubernetes, the required automation for a platform, the relevant components for a self-service, developer-friendly oriented platform, and the steps required to build secure and compliant environments.

This part has the following chapters:

- *Chapter 4, Architecting the Platform Core – Kubernetes as a Unified Layer*

- *Chapter 5, Integration, Delivery, and Deployment – Automation is Ubiquitous*

- *Chapter 6, Build for Developers and Their Self-Service*

4

Architecting the Platform Core – Kubernetes as a Unified Layer

As a platform engineering team, you need to make a critical decision about the underlying technology stack of your core platform. This decision will have a long-term impact on your organization as it will dictate the skills and resources you will need to build a platform that will support current and future self-service use cases.

Kubernetes – or **K8s** for short – is not the solution to all problems, but when building platforms, Kubernetes can build the foundation.

In this chapter, you will gain insights into what makes Kubernetes the choice for many platform engineering teams. We will explain the concept of *promise theory*, which Kubernetes is based on, and the benefits that come from the way it's been implemented.

You will get a better understanding of how to navigate the **Cloud Native Computing Foundation (CNCF)** ecosystem as it will be critical for you to pick the right projects to support you in your own platform implementation.

Once you are familiar with the benefits of Kubernetes and the ecosystem, you will learn about the considerations when defining the core layer of your platform, such as unifying infrastructure, application, and service capabilities. You will learn how to design for interoperability with your core corporate services that sit outside of your new platform and how to design for flexibility, reliability, and robustness.

As such, we will cover the following main topics in the chapter:

- Why Kubernetes plays a vital role, and why it is (not) for everyone
- Leveraging and managing Kubernetes infrastructure capabilities
- Designing for flexibility, reliability, and robustness

Why Kubernetes plays a vital role, and why it is (not) for everyone

For now, we will focus on Kubernetes, but there are other ways to provide a platform to run your workload. Besides many different flavors of Kubernetes, such as OpenShift, there are alternatives, such as Nomad, CloudFoundry, Mesos, and OpenNebula. They all have reasons for their existence, but only one has been adopted almost everywhere: Kubernetes!

Besides those platforms, you can use virtual machines or services from public cloud providers for serverless, app engines, and simple container services. In many cases, platforms utilize these services as well, when they are needed. An exclusive all-in Kubernetes strategy might take a few years longer, as it takes organizations a while to fully commit to it. However, there are two recent trends you can observe:

- Managing virtual machines from Kubernetes
- Migrating to virtual clusters and virtual machines managed by clusters to prevent cost explosions for hypervisor licenses

Kubernetes comes with a vital ecosystem and community, a wide range of use cases implemented by other organizations, and highly motivated contributors to solve the next challenges coming up with Kubernetes.

Kubernetes – a place to start, but not the endgame!

"*Kubernetes is a platform to build platforms. It's a start but not the endgame*" is a quote from Kelsey Hightower, who worked at Google when, back in 2014, Kubernetes was released to the world. However, while Kubernetes plays a vital role in building modern cloud-native platforms, this doesn't mean it's the perfect fit for everyone. Remember the product-centric approach to platform engineering? It starts with understanding the pain points of your users. Once we know the pain points, we can work on how we would implement the use cases and which technology choices to make.

Revisit the early section in *Chapter 1* called *Do you really need a platform?*, where we provided a questionnaire that helps you decide what the core of the platform will be. The answer could be Kubernetes, but it doesn't have to be. Let's start by looking into our own example use case from Financial One ACME.

Would Financial One ACME pick Kubernetes?

If we think about the use case from Financial One ACME, "*Easier access to logs in production for problem triage*", using the proposed solution doesn't necessarily require Kubernetes as the underlying platform.

If Kubernetes is not being used yet in our organization and the only thing we need is a new automation service that integrates into the different logging solutions, we may not want to propose Kubernetes as the underlying core platform. This is because it brings a new level of complexity into an organization that doesn't yet have the required experience. We could implement the solution and operate it with all the existing tools and teams; maybe we could run it alongside other tools we already have, following the same operational processes for deployment, upgrades, monitoring, and so on.

On the other hand, if there is pre-existing knowledge, or perhaps even Kubernetes is already available, then using Kubernetes as the core platform to orchestrate this new service would solve a lot of problems, such as providing the following:

- New service containers as Pods
- Automated health checks for those services
- Resiliency and scalability through concepts such as ReplicaSets and Auto-Scaling
- External access through ingress controllers and Gateway API
- Basic observability of those services through Prometheus or OpenTelemetry

However, do we really need to run our own Kubernetes cluster when we just need to deploy a simple service? The answer is no! There are alternatives, such as running the implementation using the capabilities of your preferred cloud provider:

- **Serverless**: The solution could be implemented as a set of serverless functions exposed via an API gateway. State or configuration can be stored in cloud storage services and can easily be accessed via an API.
- **Container**: If the solution is implemented in a container, that container can be lightweight and its endpoints can easily be exposed via an API gateway. There is no need for a full-fledged Kubernetes cluster that somebody needs to maintain.

This single use case for Financial One ACME may not lead us to choose Kubernetes as the core platform. However, when making this critical decision about what is to become the core of your future platform, we must also look beyond the first use case. Platform engineering will solve many more use cases by providing many self-service capabilities to the internal engineering teams in order to improve their day-to-day work.

It's a tricky and impactful decision to make, one that needs a good balance between looking forward and over-engineering. To make that decision easier, let's look into the benefits of picking Kubernetes as the core platform.

Benefits of picking Kubernetes as the core platform

To make the critical decision of picking the core of a future platform easier, let's look at why other organizations are picking Kubernetes as the core building block. Understanding those reasons, the benefits, and also the challenges should make it easier for architects to make this important decision.

Declarative desired state – promise theory

Traditional IT operations use the **obligation model**, which is when an external system instructs the target system to do certain things. This model requires a lot of logic to be put into the external system, such as an automated pipeline. A scripted pipeline, whether based on Jenkins, GitHub Actions, or other solutions, not only needs to apply changes to the target system. The pipeline also needs to deal with handling unpredicted outcomes and errors from outside the system it changes. For example, what do we do if deploying a new software version doesn't work within a certain amount of time? Should we roll it back? How would the pipeline do that?

In the Kubernetes Documentary Part 1 (`https://www.youtube.com/watch?v=BE77h7dmoQU`), Kelsey Hightower explained the promise theory model that Kubernetes follows with a great analogy. It goes something like this:

> *If you write a letter, put it in an envelope, and put the destination address and the right stamps on it, then the post office promises to deliver that letter to the destination within a certain amount of time. Whether that delivery involves trucks, trains, planes or any other form of delivery doesn't matter to the person who wrote that letter. The postal service will do whatever it takes to keep the promise of delivery. If a truck breaks down, some other truck will continue until the letter gets delivered to its final destination.*

The same principle is true for Kubernetes! In our analogy, the letter is a container image that we put into an envelope. The envelope in the Kubernetes world is a custom resource of a certain **Custom Resource Definition (CRD)**. To deliver an image, this could be a definition of a Deployment, which includes the reference to the image, the number of replicas, the namespace this image should be deployed into, and the resource requirements (CPU and memory) for the image to run correctly. Kubernetes then does everything it can to fulfill the promise of deploying that image by finding the right Kubernetes node that meets all the requirements to run the container image with the specified amount of replicas and the required CPU and memory.

Another example is an Ingress that exposes a deployed service to the outside world. Through annotations, it is possible to control the behavior of certain objects. For an Ingress, this could be the automatic creation of a TLS certificate for the domain that should be used to expose the matching services to be accessible via HTTPS. The following is an example of an Ingress object for `fund-transfer-`

`service` to expose the object via a specific domain to the outside world using the Certificate Manager – a core Kubernetes ecosystem tool – to create a valid TLS certificate from `LetsEncrypt`:

```
apiVersion: networking.k8s.io/v1
kind: Ingress
metadata:
  name: fund-transfer-service
  namespace: prod-useast-01
  annotations:
    cert-manager.io/issuer: "letsencrypt-prod"
spec:
  ingressClassName: nginx
  tls:
  - hosts:
    - fundtransfer-prod-useast-01.finone.acme
      secretName: finone-acme-tls
  rules:
  - host: fundtransfer-prod-useast-01.finone.acme
    http:
    . . .
```

There is a bit more to a fully working Ingress object description than what's shown in this manifest *[1]* example. However, this example does a good job of explaining how a definition will be translated by Kubernetes into the actual actions that one would expect – hence fulfilling the promise.

Now, the question is: "*How does all this magic work?*" To answer this, we will start by exploring the concepts of controllers and operators.

Kubernetes controllers and operators

Kubernetes controllers are essentially control loops that fulfill the promise theory of Kubernetes. In other words, controllers automate what IT admins often do manually: continuously observe a system's current state, compare it with what we expect the system to look like, and execute remedial actions to keep the system running!

A core task of Kubernetes controllers is therefore *continuous reconciliation*. This continuous activity allows it to enforce the desired state, for example, making sure that the *desired state* expressed in the *Ingress definition* example from earlier matches the *current state*. If either the desired state or the current state changes, it means they are *out of sync*. The controller then tries to synchronize the two states by making changes to the managed object until the current state matches the desired state again!

The following illustration shows how a controller watches the *desired state* (expressed through manifests and stored in etcd), compares it with the *current state* (the state persisted in etcd), and manages the *managed objects* (e.g., Ingress, Deployments, SSL certificates, and so on):

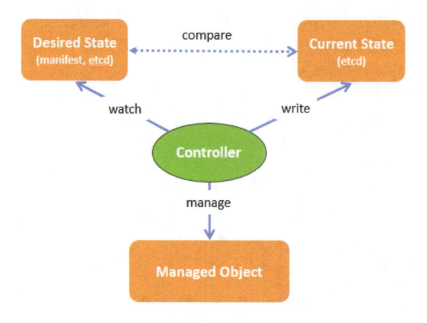

Figure 4.1: Reconciliation and self-healing by design through Kubernetes controllers

This figure already shows the core concepts and power of controllers, highlighting how the automated reconciliation loop ensures automated self-healing by design. However, controllers fulfill other functions as well: observing cluster and node health, enforcing resource limits, running jobs on a schedule, processing life cycle events, and managing deployment rollouts and rollbacks.

Now, let's discuss Kubernetes operators. Operators are a subcategory of controllers and typically focus on a specific domain. A good example is the OpenTelemetry operator, which manages OpenTelemetry Collectors and the auto-instrumentation of workloads. This operator uses the same reconciliation loop to ensure that the desired configuration for OpenTelemetry is always applied. If the configuration is changed or if there is a problem with the current OpenTelemetry Collector or instrumentation, the operator will do its best to keep the promise of ensuring that the desired state is the actual state. To learn more, visit the OpenTelemetry website at `https://opentelemetry.io/docs/kubernetes/operator/`.

Other use cases for operators typically relate to managing and automating core services and applications such as databases, storage, service meshes, backup and restore, CI/CD, and messaging systems.

If you want to learn more about controllers and operators, have a look at the CNCF Operator Working Group and their white paper at `https://github.com/cncf/tag-app-delivery/tree/main/operator-wg`. Another excellent overview can be found on the *Kong Blog* post titled *What's the Difference: Kubernetes Controllers vs Operators*. This blog also lists great examples of controllers and operators: `https://konghq.com/blog/learning-center/kubernetes-controllers-vs-operators`

Now that we know more about controllers and operators, let's have a look at how they also ensure built-in resiliency for all Kubernetes components and deployments!

Built-in resilience driven by probes

Controllers continuously validate that our system is in its desired state by observing the health of the Kubernetes cluster, along with its nodes and all deployed Pods. If one of the observed components is not healthy, the system tries to bring it back into a healthy state through certain automated actions. Take Pods, for example. If Pods are no longer healthy, they eventually get restarted to ensure the overall system's resiliency. Restarting components is also often the default action an IT admin would execute following the "*Let's try to turn it off and on again and see what happens!*" approach.

Just like IT admins who probably won't just turn things on and off at random, Kubernetes follows a more sophisticated approach to ensuring the resiliency of our Kubernetes clusters, nodes, and workloads.

Kubelet – a core component of Kubernetes – continuously observes the life cycle and the health state of Pods using several types of probes: startup, readiness, and liveness. The following illustration shows the different health states a pod can be in depending on the results of startup, readiness, and liveness probe checks:

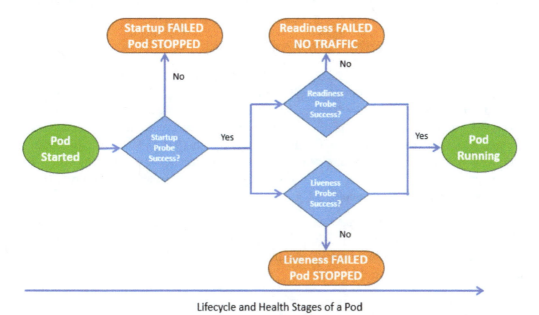

Lifecycle and Health Stages of a Pod

Figure 4.2: Kubelet determining the health status of Pods using probes

Once Pods are no longer healthy, Kubernetes will try to restart Pods and bring the Pods back to a healthy state. There are a lot of different settings for both the evaluation of the probe results and the restart policies, which you must familiarize yourself with to fully take advantage of the built-in resiliency of Kubernetes. All those settings are declared on your Deployment and Pod definitions.

If you want to learn more about how Kubelet manages the different probes and see some best practices, we can recommend checking out blog posts such as the one from Roman Belshevits on liveness probes: `https://dev.to/otomato_io/liveness-probes-feel-the-pulse-of-the-app-133e`

Another great resource is the official Kubernetes documentation on configuring liveness, readiness, and startup probes: `https://kubernetes.io/docs/tasks/configure-pod-container/configure-liveness-readiness-startup-probes/`

> **Health probes are only valid within Kubernetes**
>
> It's important to understand that all these health checks are only done within the Kubernetes cluster and don't tell us whether a service that is exposed via an Ingress to our end users is also considered healthy from the end users' perspective. A best practice is to additionally check health and availability from an "outside-in" perspective as an external control. For example, you can use synthetic tests to validate that all exposed endpoints are reachable and return successful responses.

Now that we have learned about built-in resiliency for Pods, how about more complex constructs, such as applications that are typically made up of several different Pods and other objects, such as Ingress and storage?

Workload and application life cycle orchestration

As we have learned, Kubernetes provides built-in orchestration of the life cycle of Pods, as explained in the previous section. However, business applications that we deploy on Kubernetes typically have multiple dependent Pods and workloads that make up the application. Take our Financial One ACME as an example: the financial services applications deployed to support its customers contain multiple components, such as a frontend, a backend, caches, databases, and Ingress. Unfortunately, Kubernetes doesn't have the concept of applications. While there are several initiatives and working groups to define an application, we currently have to rely on other approaches for managing applications, which are composites of multiple components.

In the *Batching changes to combat dependencies* section in *Chapter 5*, we will learn about tools such as Crossplane. Crossplane allows you to define so-called composites, which make it easy for application owners to define individual components of an application and then deploy individual instances, as shown in the following example:

```
apiVersion: composites.financialone.acme/v1alpha1
kind: FinancialBackend
metadata:
  name: tenantABC-eu-west
spec:
  service-versions:
    fund-transfer: 2.34.3
```

```
    account-info: 1.17.0
redis-cache:
  version: 7.4.2
  name: transfer-cache
  size: medium
database:
  size: large
  name: accounts-db
region: "eu-west"
ingress:
  url: „https://tenantABC-eu-west.financialone.acme"
```

Crossplane provides application and infrastructure orchestration and uses the operator pattern to continuously ensure that every application instance – as defined in the composite – is running as expected.

Another tool we will learn more about in *Chapter 5* is the **Keptn** CNCF project. Keptn provides automated application-aware life cycle orchestration and observability. It gives you the option to declaratively define pre- and post-deployment checks (validate dependencies, run tests, evaluate health, enforce SLO-based quality gates, and so on) without having to write your own Kubernetes operator to implement those actions. Keptn also provides automated deployment observability to better understand how many deployments happen, how many are successful, and where and why they fail by emitting OpenTelemetry traces and metrics for easier troubleshooting and reporting.

Kubernetes provides a lot of the building blocks for building resilient systems. While you can write your own operators to expand this to your own problem domain, you can also use existing CNCF tools such as Crossplane or Keptn as they provide an easier declarative way to apply the concept of promise theory to more complex applications and infrastructure compositions.

Restarting components is one way of ensuring resiliency, but there are more. Let's have a look at auto-scaling, which solves another critical problem in dynamic environments!

Auto-scaling clusters and workloads

In most industries, the load expected on a system is not equally distributed across every day of the year. There is always some type of seasonality: retail gets spikes on Black Friday and Cyber Monday, tax services get spikes on tax day, and finance often spikes when paychecks are coming. The same is true for our own Financial One ACME customers. As a financial services organization, there is always some basic traffic from end users, but there will be spikes at the beginning and end of the month.

Kubernetes provides several ways to scale application workloads: manually (e.g., setting ReplicaSets) or automatically through tools such as **Horizontal Pod Autoscaler (HPA)**, **Vertical Pod Autoscaler (VPA)**, or **Kubernetes Event Driven Autoscaler (KEDA)**. Those scaling options allow you to scale when your workloads run low on CPU or memory, when applications see a spike in incoming traffic, or when response time is starting to increase!

Besides workloads, you can and most likely have to also scale the size of your clusters and nodes through tools such as **Cluster Autoscaler** (**CA**) or **Karpenter** *[2]*, or through options available via your managed Kubernetes cloud vendor.

As a platform engineering team, you need to make yourself familiar with all the different options but also be aware of all the considerations:

- **Setting limits**: Don't allow applications to scale endlessly. You have options to enforce maximum limits per application, workload, namespaces, and more.

- **Cost control**: Auto-scaling is great but has a price tag. Make sure to report costs to the application owners.

- **Scale down**: Scaling up is easy! Make sure to also define indicators for when to scale down. This will keep costs under control.

To learn more about them, please review the documentation: `https://kubernetes.io/docs/concepts/workloads/autoscaling/`

Now that we have learned about the options for scaling within a Kubernetes environment, how about scaling out to other Kubernetes clusters?

Declare once – run anywhere (in theory)

The promise of Kubernetes as an open standard is that any declared state (Ingress, workloads, secrets, storage, network, etc.) will behave the same whether you run it on a single cluster or on multiple clusters to meet certain requirements, such as the separation of stages (dev, staging, and production) or the separation of regions (US, Europe, and Asia).

The same promise holds true in theory whether you operate your own Kubernetes cluster, use OpenShift, or use a managed Kubernetes service from one of the cloud vendors. What does *in theory* mean here? There are some specific technical differences between the different offerings you need to take into consideration. Depending on the offering, networking or storage may act slightly differently because the underlying implementation depends on the cloud vendor. Certain offerings will also come with specific versions of core Kubernetes services, services meshes, and operators that come with a managed installation. Some offerings require you to use vendor-specific annotations to configure the behavior of certain services. That's why applying the same declarative state definition across different vendors will, *in theory*, work – in practice, you have to consider certain small differences that require some vendor-specific configuration!

As the technical details and differences are constantly changing, it wouldn't make sense to provide a current side-by-side comparison as part of this book. What you must understand is that while, in theory, you can take any Kubernetes object and deploy it on any flavor of Kubernetes, the outcome and behavior might be slightly different depending on where you deploy it. That's why we suggest doing some technical research on the chosen target Kubernetes offering and how it differs from other offerings in case you want to go for multi-cloud/multi-Kubernetes, because the same Kubernetes objects might behave slightly differently!

The good news is that the global community is working to solve this problem by providing better guidance and tools to make the *declare once – run anywhere* promise a reality.

We've looked at a lot of the benefits of picking Kubernetes as the core platform. However, there is another good reason why Kubernetes has seen such great adoption over the past 10 years since its first release: the global community and the CNCF!

Global community and CNCF

Kubernetes was announced by Google in June 2014, and version 1.0 was released on July 21, 2015. Google then worked with the Linux Foundation and formed the CNCF with Kubernetes as its initial project! Since then, the community and the projects have taken the world by storm!

10 years later (at the time of writing this book), the CNCF has 188 projects, 244,000 contributors, 16.6 million contributions, and members in 193 countries worldwide. Many presentations that introduce Kubernetes and the CNCF often start by showing the CNCF landscape: `https://landscape.cncf.io/`

While the landscape is impressive, it has also been the source of many memes about how hard and complex it is to navigate the landscape of all projects this global community is working on. However, don't be scared. The global CNCF community is part of the Linux Foundation and has the mission to provide support, oversight, and direction for fast-growing cloud-native projects, including Kubernetes, Envoy, and Prometheus.

Here are a few things you should be aware of because they will help you navigate the ever-growing list of CNCF projects in the project landscape:

- **Project status**: CNCF actively tracks the status and activity of every project. The number of contributors and adopters, as well as how active development is for a project, are good indicators of whether you should look closer at a project. Projects that are stale, only have a single maintainer, or hardly have any adopters might not be of any use if you are deciding on tools that will help you for the long term in your platform.

- **Maturity level**: The CNCF also specifies a maturity level of sandbox, incubating, or graduated, which corresponds to the Innovators, Early Adopters, and Early Majority tiers of the *Crossing the Chasm* diagram (`https://en.wikipedia.org/wiki/Crossing_the_Chasm`). Graduated projects have been adopted widely across various industries and are a safe choice for the use cases they support. Incubating projects have crossed over from a technical playground to seeing good adoption with growing numbers of a diverse set of maintainers. To learn more about the criteria for CNCF maturity and to see who is at which level, check out the official site at `https://www.cncf.io/project-metrics/`.

- **Adopters**: Every CNCF project tries to increase and track adoption. One way of doing this is by making organizations that actively adopt a project add themselves to the ADOPTERS.md file, which every CNCF project typically has in its GitHub repository. If you decide to adopt one of those projects, we encourage you to also add your name to the list of adopters by opening up a pull request. This helps the project and will help other organizations decide whether this is a project worth pursuing!

> **Kubernetes is vital because of its community**
>
> While Kubernetes has a strong technology base, it is really the community and the ecosystem that was built over the past 10+ years that makes Kubernetes a viable option for platform engineers to use as their core platform.

We have now learned more about what Kelsey Hightower meant when he said: "*Kubernetes is a platform to build platforms. It's a start but not the endgame.*" There are many benefits of picking Kubernetes as the core platform, especially as it is built on the concept of *promise theory*. Kubernetes provides automated resiliency, scaling, and life cycle management of components. The ever-growing community provides solutions to many common problems through hundreds of open source CNCF projects that every organization can use.

While we often focus on the benefit of Kubernetes for deploying and orchestrating applications, let's have a look at how we can use Kubernetes to lift our infrastructure capabilities into our future platform!

Leveraging and managing Kubernetes Infrastructure Capabilities

Back in *Chapter 2*, you were introduced to the Platform Reference Components model and the capability plane. When we are writing about lifting infrastructure capabilities to Kubernetes, the end user becomes aware of those capabilities when using the platform. We must differentiate between resources that need to be integrated with Kubernetes and those configured by specifications deployed to Kubernetes and manipulated or created new resources outside of the cluster. In the following figure, you can find examples in the resource integration and network section that require a solid integration; otherwise, they actively prevent a useful and functioning platform.

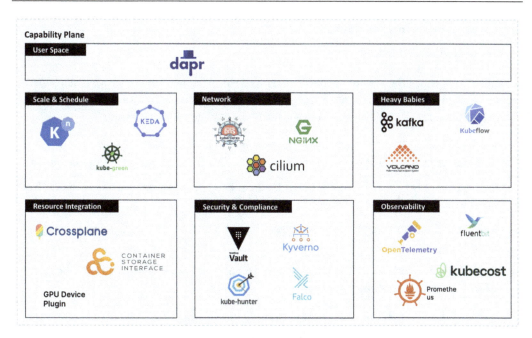

Figure 4.3: Capability plane with example tools

Integrating infrastructure resources

We will discuss the basic components and design decisions you must make for the platform's underlying technologies. Firstly, due to Kubernetes' power, you are more flexible in your tooling and can extend it as needed. This is especially helpful when you're adjusting the platform's capabilities for different use cases.

Container storage interface

The **Container Storage Interface (CSI)** provides access to the storage technology that is attached to the cluster or the nodes running the cluster. In the CSI developer documentation *[3]*, you can find a driver for almost every storage provider. The list contains cloud provider drivers such as AWS **Elastic Block Storage (EBS)**, software-defined storage such as Ceph, or a connector for commercial solutions such as NetApp. In addition, the CSI driver also supports tool-specific drivers such as cert-manager and HashiCorp Vault. In short, the CSI is vital to any data that should live longer than the container it belongs to and is not stored in a database, or is needed for database storage.

The installation of the driver depends on the infrastructure and storage technology. For a cloud provider, for example, you usually require the following:

- A service account or permission policies
- Configuration for startup taints and tolerations

- Pre-installed external snapshotter

- The driver installation itself

Due to their complexity, these components are deployed with Helm or other package management solutions. Sometimes, they require more privileges on the node, which can be a security concern when designing the platform. You will also need to consider how storage will be accessed:

- **ReadWriteOnce (RWO)**: One Pod claims a portion of the available storage, which it can read from and write to, while other Pods cannot access it unless the original Pod releases the storage.

- **ReadWriteMany (RWM)**: Multiple Pods can claim one portion of the available storage. They can read and write to it, and share that storage with others.

- **ReadOnlyMany (ROM)**: Multiple Pods can claim one portion of the storage, but only to read from it.

- **ReadWriteOncePod (RWOP)**: Can be claimed by only one Pod; no other Pod can take it, and it allows read and write operations.

The overview of the CSI driver provides further information on which access modes are supported. As there is no one-size-fits-all solution, you have to make your options transparent to the user and explain how to use them.

The core elements to know and to understand for your users are `StorageClass`, `PersistentVolume`, and `PersistentVolumeClaim`. When a Pod/Deployment requires a volume, `StorageClass` will trigger the creation of a new volume. In case a Pod/Deployment has already claimed a volume once, and it didn't get destroyed, the Kubernetes control plane will re-assign the volume to the Pod/Deployment.

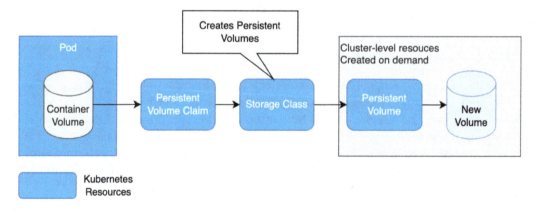

Figure 4.4: Dynamically provisioning new persistent volumes

You define `StorageClass` as the platform team, ideally collaborating with your storage experts. The following example highlights the common definition of `StorageClass`. There is plenty of room to make a mistake in the configuration; for example, setting no `reclaimPolicy` will, by default, delete the later created and attached volume. However, `StorageClass` supports you in the dynamic creation of volumes for user-requested `PersistantVolumeClaim` and is therefore a strong enabler of self-service:

```
apiVersion: storage.k8s.io/v1
kind: StorageClass
metadata:
  name: ultra-fast
  annotations:
    storageclass.kubernetes.io/is-default-class: "false"
provisioner: csi-driver.example-vendor.example
reclaimPolicy: Retain # default value is Delete
allowVolumeExpansion: true
mountOptions:
  - discard
volumeBindingMode: WaitForFirstConsumer
parameters:
  guaranteedReadWriteLatency: "true" # provider-specific
```

The downside of this is that you have to consider human error, which can take down your platform by filling up the storage until the system freezes.

CSI challenges

The definition of a CSI is a fairly new approach, but the underlying technologies of storage management and software-defined storage are way older than Kubernetes. We can also discover this in many CSI drivers, which are nothing but a shim wrapping legacy code. For less flexible and scalable clusters, this might not be a problem, but in environments where you have a lot of action going on, you don't want to have a CSI in your system that becomes the bottleneck. Some CSIs even have restrictions and limitations to prevent their failure at scale. To be fair, we usually see this with on-premises installations and some old storage technologies. With those, we can add the **Logical Unit Number** (**LUN**) presentation and connection limits on top of the things to consider. The LUN is for the Pods to make requests from storage space and retrieve data. There are limits on how many connections a physical server can have to the storage. Again, this is important when you manage your own storage and SANs.

Why are some CSIs so poor? The CSI does more than provide storage capacity. It communicates with the storage provider, promises the availability of the demanded capacity, and waits until the RAID controller, backup, and snapshot mechanisms are ready. In large-scale storage systems, we get even more: we can find storage capacity allocation and optimization procedures.

To overcome such issues, you need to evaluate the storage, especially the storage drivers, of their cloud-native storage capabilities. This will require large-scale performance tests and a ridiculous number of created volumes. The CSI driver shouldn't do any cut-downs or performance decrease and the created volume should be in a millisecond area.

Furthermore, ensure that the driver allows cross-storage and cross-cloud/infrastructure migration; provides synchronous and asynchronous replication between those different infrastructures; provides feature parity across sites; and supports local storage types if needed, such as for edge scenarios.

A good CSI will enable your platform to operate anywhere and to support a wide range of use cases.

Container network interface

The **Container Network Interface (CNI)** can become a platform's most relevant component, but it is also its most underrated one. For many projects we have seen, some platform teams don't care what they use as CNI, nor do they heavily utilize network policies, encryption, or fine-grained network configurations. Thanks to its simple abstraction of the network layer, it's not overwhelming when getting started. Yet, on the other hand, there are many use cases where the most crucial component is the CNI. I have even seen projects fail because of the dynamic nature of a CNI that didn't play along with the very traditional and legacy approach of implementing networks.

A CNI always requires a dedicated implementation because it is a set of specifications and libraries for writing plugins to configure network interfaces. The CNI concerns itself only with the network connectivity of Linux containers and removing allocated resources when the container is deleted. Due to this focus, CNIs have a wide range of support, and the specifications are simple to implement.

Therefore, we architects should never treat the CNI as *"just another object in Kubernetes."* We have to evaluate and introduce it. Some CNIs are made for easy maintenance and a solid but simple set of features. For example, if the primary focus is on layer 3, consider Flannel. Other CNIs, such as Calico, are a rock-solid choice with a rich feature set; Cilium introduced the usage of eBPF to provide even faster and more secure networking. If it's still difficult to choose between those options because you may have additional requirements such as providing different levels of networks, then the community still has an answer for you: Multus. Take your time and discover your options. There are dozens of CNIs.

The CNI can have serious effects on your platforms:

- **Security**: CNIs can provide different capabilities for network policies to achieve fine-grained control over your network. They can have additional encryption features, integrations into identity and access management systems, and detailed observability.

- **Scalability**: The larger the cluster gets, the more communication happens throughout the network. The CNI must support your growth target and stay efficient even with complex routing and chatting across the wires.

- **Performance**: How fast and direct can Pod-to-Pod communication be? How much complexity does the CNI introduce? How efficiently can it handle communication? Can it deal with many complex network policies?

- **Operability**: High-level CNIs are not very inversive and simple to maintain. Powerful CNIs can, in theory, be replaced as long as they adhere to the CNI specification, but each comes with its own set of features, which are often not replaceable.

Be aware that not every CNI supports all of those features. Some, for example, do not even provide network policy support. Other CNIs are cloud-provider-specific integrating with just one cloud provider and they do enable some cloud provider-centric capabilities.

Architectural challenges – CNI chaining and multiple CNIs

Platforms are predestined for CNI chaining. CNI chaining introduces the sequential usage of multiple CNIs. The order in which CNIs are taken and for what purpose is defined in the `/etc/cni/net.d` directory and handled by the kubelet. This allows the platform team to handle one part of the network and provide a guarded approach, while platform users can freely configure their network at a higher level. For example, a platform user can access Antrea as a CNI to configure their networking to some extent. They can also apply network policies and egress configurations to prevent their application from chatting with everyone. On the other side of the platform, the platform engineering team will manage, via Cilium, the global cross-cluster communication, as well as the network encryption, to enforce security best practices. In addition, the networking data made visible by Cilium is made available to the operations and security teams. Where those use cases are most suitable is in the interaction with the cloud providers' own CNIs. They often enable better integration between the platform and the cloud but lack many advanced features on the other side.

Another approach to be evaluated would be to assign a Pod multiple network interfaces via Multus or CNI-Genie. Normally, a Pod has just one interface, but with Multus, for example, this could be multiple network interfaces. When does this become relevant? The following instances are examples when it becomes relevant:

- Separating control and operational data from application data

- Providing flexible network options for an extremely heterogeneous workload

- Taking multi-tenancy to another level by assigning completely different networks for each tenant

- Supporting unusual network protocols and connections, such as in edge scenarios or telco environments

- **Network Function Virtualization** (**NFV**), which requires multiple networks due to its complexity

The Multus CNI is a kind of meta-plugin on the node and sits between the actual CNIs and the Pod network interfaces, as shown in the following illustration. It attaches the different network interfaces to the Pod on one side and handles the connection to the anticipated CNIs for the right network interface on the other side.

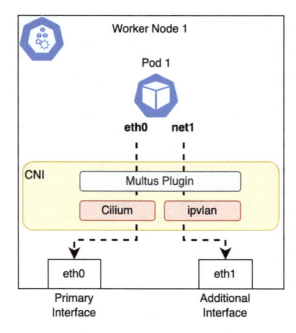

Figure 4.5: Multus meta-CNI plugin

Both approaches must be evaluated well. They have a significant impact on your network complexity, performance, and security.

Providing different CPU architectures

Kubernetes supports multiple CPU architectures: AMD64, ARM64, 386, ARM, ppc64le, and even mainframes with s390x. Many clusters today run on AMD64, but at the time of writing, a strong interest in ARM64 is causing a shift. This discussion is primarily about saving costs and gaining a little bit of extra performance while reducing the total power consumption. At least on paper, it is a win-win-win situation. Not only is the ARM64 a possible change in the infrastructure, the open source architecture project RISC-V is gaining speed and is the first cloud provider to create RISC-V offerings *[4]*.

As platform engineers, we can enable those migrations and changes. A Kubernetes cluster can run multiple architectures simultaneously—not on the same node but with different groups of nodes. Remember that this change also requires an adjustment in the container build. With some software components, you can do a multi-architecture build; with others, it might require adjustments before the container for a different architecture can be created.

To select the architecture on which a Deployment should be delivered, you just need to add a nodeSelector like this to the Spec section of the Deployment YAML file:

```
nodeSelector:
        kubernetes.io/arch: arm64
```

An upcoming alternative to provide different container images is to compile the software as a **WebAssembly (Wasm)** container.

Wasm runtime

The usage of Wasm as an alternative container format has increased drastically in the last year. Wasm is a binary instruction format for a stack-based virtual machine. Think of it as an intermediate layer between various programming languages and many different execution environments. You can take code written in over 30 different languages, compile it into a `*.wasm` file, and then execute that file on any Wasm runtime.

The name *WebAssembly*, however, is misleading. Initially designed to make code run quickly on the web, today, it can run anywhere. Why should we use it? Wasm is secure and sandboxed by default. In a Wasm container, there is no operating system or anything else except the binary compiled code. This means there is nothing to break into or claim the context or service account from. Wasm also has an incredibly fast startup time where the limits are set by the runtime rather than the module. Some runtimes claim to be as fast as around 50 ms. In comparison, your brain requires >110 ms to recognize whether something passes in front of your eyes. Furthermore, a Wasm container size is around 0.5 MB – 1.5 MB, whereas a very slim container can be around 5 MB. However, what we can see on the market is usually image sizes in the range of 300 MB – 600 MB, 1 GB – 3 GB, or even sometimes above 10 GB. With this reduced image size, a Wasm container also has a drastically reduced storage footprint. Wasm is also hardware-independent. The exact same image can run anywhere, as long as you have a Wasm runtime available.

In the context of Kubernetes, the OCI and CRI runtimes support Wasm. This means you can run a Wasm container alongside a regular container. As you can see in the following figure, no further changes are required. The Wasm app image is stored at the node level and executed by the layers above.

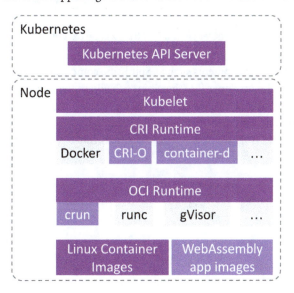

Figure 4.6: Wasm on a Kubernetes node

To make Wasm executable and available in your platform, you have to specify a runtime class and define its usage at the Deployment/Pod level. In the following example, you can see the specification for `crun` as `RuntimeClass` on the left side, and a `Pod` definition where, for `spec.runtimeClassName`, we assign `crun` on the right side. For `crun`, we also have to add an annotation to inform `crun` that this `Pod` has a Wasm image:

```apiVersion: node.k8s.io/v1`  `kind: RuntimeClass`  `metadata:` ` name: crun` `scheduling:` ` nodeSelector:` ` runtime: crun` `handler: crun```	```apiVersion: v1`  `kind: Pod`  `metadata:` ` name: wasm-demo-app`  ` annotations:`  ` module.wasm.image/variant: compat` `spec:`  ` runtimeClassName: crun`  ` containers:` ` name: wasm-demo-app` ` image:` ` docker.io/cr7258/wasm-demo-app:v1```

Table 4.1: Definition of a RuntimeClass and a Pod that will be executed in the Wasm runtime

As an alternative to that, new developments such as SpinKube come with a whole set of tools to utilize Kubernetes resources in the best manner *[5]*. That approach allows the integration of the development experience into the deployment and the execution of the Wasm containerized app. The following image shows the workflow and how the different components work together. It makes the development process straightforward and brings a fast but robust new runtime environment to the platform.

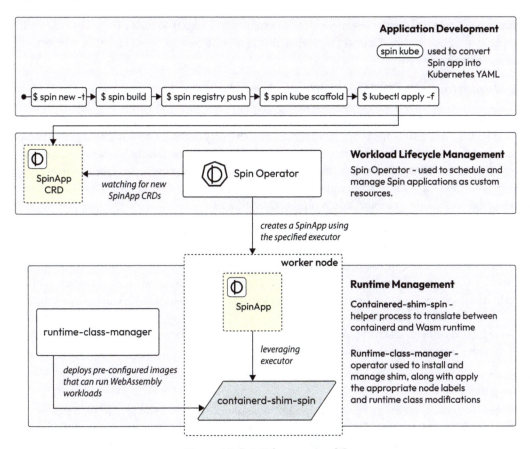

Figure 4.7: SpinKube overview [6]

However, is Wasm really a technology that the industry has adopted? Here are just some examples and use cases showing its adoption:

- **Interoperability**: Figma runs as Wasm on any computer

- **Plugin system**: You can write extensions in Wasm for Envoy

- **Sandboxing**: A Firefox or Chrome browser can run Wasm in a sandboxed environment to protect your system

- **Blockchain**: CosmWasm or the ICP uses versions of Wasm to run applications

- **Container**: WasmCloud has a different concept to execute containers, or SpinKube, a Wasm runtime with a CLI for a simple development process

- **Serverless platforms**: Cloudflare Workers or Fermyon Cloud run your Wasmized app

Wasm is not a container replacement yet. It is an evolutionary step that suits a few cases very well, but it lacks adoption and has issues in the debugging process. However, it is a matter of time before these obstacles are solved.

Enable platforms for GPU utilization

Similar to the support for different CPU architectures or the extension of the container runtime to include Wasm, GPU enablement for users has become extremely relevant lately. A device plugin must be installed that is specific to the type of GPU, such as AMD, Intel, or Nvidia. This exposes custom schedulable resources such as `nvidia.com/gpu` to Kubernetes and its users. Also, we are not experts in GPUs and their capabilities; from a platform engineering perspective, the different providers have developed diverse feature sets and extensions for their plugins.

I highly recommend developing or finding experts for this field if your user requires GPUs. The field of AI and LLMs is undergoing rapid expansion. Hardware and software providers come up with new tools, systems, and approaches monthly to take advantage of those technologies. Any scenario has its own very specific demands. Training a model requires a tremendous amount of data, GPUs, storage, and memory. Fine-tuning a model comes primarily down to how large a model you want to train. A seven-billion-parameter model can fit into a 14 GB VRAM, but increasing the precision can increase its size easily to 24 GB or more. Lastly, providing an inference engine to serve an LLM for users requires a lot of network communication.

> **Important note**
> Inferencing means sending a prompt to an LLM. Most people believe that the LLM then creates the story like a human would. However, what is really happening is that after every word, the LLM has to send the whole prompt, including the new words, to the LLM again, so it can decide on the next word added to the sentence.

As a rule of thumb, for the GPU VRAMs (Video RAM is the GPU's memory) needed, you double the model's size. Here are some more examples:

- Llama-2-70b requires 2 * 70 GB = 140 GB VRAM
- Falcon-40b requires 2 * 40 GB = 80 GB VRAM
- MPT-30b requires 2 * 30 GB = 60 GB VRAM
- bigcode/starcoder requires 2 * 15.5 = 31 GB VRAM

Therefore, collaboration between the platform and the machine learning team is required. Depending on the models they want to use, your chosen hardware may no longer fit the requirements.

Architectural challenges

While the integration of GPUs is very straightforward, they have a few elements that need to be discussed; you should be aware of the following:

- **GPU costs**: Cheap GPUs, with low VRAM and computational power, might not fit your use cases. Also, you might not use a GPU 24/7, but you should think about more dynamic and flexible possibilities. However, owning GPUs can be cheaper in the long run as compared to their regular CPU counterparts if your use case requires GPU computational power.

- **Privileged rights**: Many machine learning tools require further customization and tweaking, especially for the rights they demand.

- **End user requirements**: Besides the model sizes, what the data scientists and machine learning engineers want to do with the model depends very much on the actual use case they want to implement. Any minor change in the approach can make the platform unusable. This must be considered in the architecture for such a platform and to provide the most resources and greatest scaling flexibility possible.

- **External models pulled to the cluster**: As with containers, it is a common practice to pull models from pages such as HuggingFace. You might consider this in the network of a platform supporting machine learning activities.

Creating platforms that suit machine learning and LLM operations requires a new level of optimization. Non-running GPUs are a waste of money. Poorly used GPUs are a waste of money. However, there is more we have to ensure from an infrastructure perspective: data protection, platform security, and specialized observability for the models. In my opinion, machine learning and LLMs are an exciting use case and offer a playground for platform engineers.

Solution space

To optimize GPU usage, there are some approaches available:

- **Multi-Process Server (MPS)**
- **Multi-Instance GPU (MIG)**
- **Time-slicing/sharing**

Depending on the GPU driver, you will find a full range of possibilities, as you do with Nvidia. Other GPU drivers might not be mature enough or feature-rich yet, but of course, you should evaluate this frequently.

Time-slicing is the worst option to take. Although it is better than nothing, MPS would be at least twice as efficient as the time-slice approach. However, MPS has one major drawback: processes are not strictly isolated, which leads to correlated failures between the slices. This is where MIG comes into the picture. It provides good process isolation and a static partitioning of the GPU. Static on a super dynamic, scalable, and anytime adjustable Kubernetes cluster? Yes, because machine learning training will not run just for a few seconds or minutes *[7]*.

The following figure shows the GPU's memory partitions at a high level, to which different workloads can be assigned.

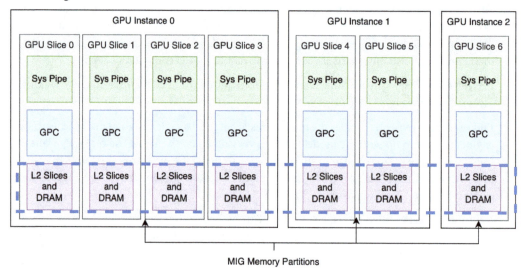

Figure 4.8: Example of splitting a GPU into three GPU instances

Those GPU instances can be used either by a single pod, a pod with one container running multiple processes (not ideal), or by using something such as CUDA from Nvidia. CUDA is an MPS, so you can combine the different approaches, as shown in the following diagram:

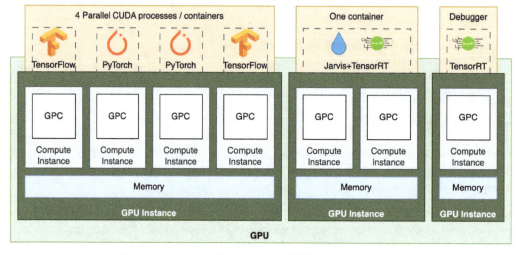

Figure 4.9: Example of using three GPU instances in parallel

From here, we can take a look at the research and experiment corner where you can combine Kubernetes **Dynamic Resource Allocation (DRA)** with MIG. It provides an approach to being flexible in the assignment while ensuring the resources for the deployments. The DRA was introduced with Kubernetes `v1.26` and is still in Alpha, therefore breaking API changes are likely with every release. There are some interesting articles and talks about it. Depending on when you are reading this, it might be out of date *[8]*.

Enable cluster scalability

As mentioned earlier in this chapter, the ability of a Kubernetes cluster to adjust its scale is beneficial for different demands, from resiliency to growing with the workload to fallback re-initiation, to providing the highest availability across data centers and availability zones. At its core, we differentiate between the horizontal and vertical autoscaler, which targets the Pods, and the horizontal CA, which adjusts the number of nodes.

As with many capabilities, Kubernetes provides the specification but expects someone else to implement it. This at least applies to the VPA and the CA, which require at least a metrics server running at the cluster level.

However, the HPA is feature-rich and allows metric-based scaling. Look at the following example of an HPA. With `stabilizationWindowSeconds`, we can also tell Kubernetes to wait on previous actions to prevent flapping. Flapping is defined as follows according to the Kubernetes documentation:

> *When managing the scale of a group of replicas using the HorizontalPodAutoscaler,*
> *it is possible that the number of replicas keeps fluctuating frequently due to the*
> *dynamic nature of the metrics evaluated. This is sometimes referred to as thrashing,*
> *or flapping. It's similar to the concept of hysteresis in cybernetics.*

We can take the following configuration for an HPA. It looks simple; you can see that based on the policies, the behavior can change drastically. For example, when reducing 10% of the Pods while having a large deployment with hundreds of replicas, we want to be very careful. The shown scaling-down configuration will prevent deleting more than two Pods at the same time:

```
apiVersion: autoscaling/v2
kind: HorizontalPodAutoscaler
spec:
  maxReplicas: 7          #maximum replicas to scale up to
  minReplicas: 3          #expected minimum replicas
scaleTargetRef: ...
  targetCPUUtilizationPercentage: 75     #metric to react on
  behavior:
    scaleDown:
      stabilizationWindowSeconds: 300      #wait 300 sec
    policies:
```

```
  - type: Percent
    value: 10                      #reducing 10% of Pods
    periodSeconds: 60              #per minute
  - type: Pods
    value: 5                       #reducing not more
    periodSeconds: 60              #than 2 Pods per min
  selectPolicy: Min
scaleUp:
  stabilizationWindowSeconds: 0
  policies:
  - type: Percent
    value: 75                      #increase the Pod by 75%
    periodSeconds: 15              #every 15 seconds
  selectPolicy: Max
```

Issues with VPA, HPA, and CA

There are some constraints that you have to consider for the autoscaler family:

- **HPA**:

  - You have to set CPU and memory limits and requests on Pods correctly to prevent resource waste or frequently terminated Pods.

  - When HPA hits the limit of the available nodes, it can't schedule more Pods. However, it might utilize all available resources, which could lead to issues.

- **VPA**:

  - VPA and HPA shouldn't be used for scaling based on the same metric. For example, while both can use CPU utilization to trigger a scale-up, HPA deploys more Pods, while VPA increases the CPU limits on existing Pods, which can lead to excessive scaling based on the same metric. Therefore, if using both VPA and HPA, one should use CPU for one and, for instance, memory for the other.

  - VPA might recommend using more resources than available within the cluster or the node. This causes the Pod to become unschedulable.

- **CA**:

  - The CA scales are based on the requests and limits of the Pods. This can cause a lot of unused resources, poor utilization, and high costs.

  - When a CA triggers a scaling command to the cloud provider, this might take minutes to provide new nodes for the cluster. During this time, the application performance is degraded. In the worst case, it becomes unservable.

Solution space

As platform engineers, we want to ensure that the user can define scaling behavior without putting the platform at risk. Utilizing HPA, VPA, and CA requires perfect configuration and control, guardrails provided by a policy engine, and close monitoring. It becomes mission-critical to control cluster scaling and descaling while enabling in-namespace autoscaling for your users.

Managing scaling your Kubernetes cluster on cloud providers requires you to look into the CA and the different available cloud integrations for it. Besides, if you use the **Cluster API (CAPI)**, you can also build on its capability for cluster autoscaling.

Network capabilities and extensions

Now, let's look at the final resource integration of Kubernetes and the underlying infrastructure. To do this, we will start within the cluster networking mechanisms and work down to the DNS and load balancing. DNS and load balancing can happen within the cluster and in coordination with the infrastructure that Kubernetes runs on.

Ingress – the old way

Ingress is the old definition of how an end user request from outside the cluster is routed into the system and toward the application that is exposed. For almost a decade, it was the way to go to define incoming network traffic. The ingress is usually represented by an ingress controller such as NGINX, HAProxy, or Envoy, to name a few. Those inherently take the routing rules defined as a standard resource from Kubernetes and manage the rest of it. As you can see in the following figure, from there on, the traffic is redirected to the right service, which forwards it to the Pod. Physically, the traffic will go from the network interface to the ingress controller to the Pod, but as good an orchestrator as Kubernetes is, there are some logical steps in between.

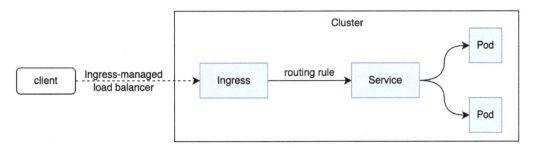

Figure 4.10: Ingress controller (source: Kubernetes docs)

While it scales and is robust for some of the largest deployments out there, it has some downsides:

- The Ingress API only supports TLS termination and simple content-based request routing of HTTP traffic

- It is limited in its available syntax and kept very simple

- It requires annotations for extensibility

- It reduces portability because every implementation has its own approach to do so

Doing a multi-tenant cluster is more challenging because Ingress is usually at the cluster level and has a poor permission model. Also, it supports namespaced configurations, but it is not suitable for multi-teams and shared load-balancing infrastructure.

Some of the beauty of the Ingress approach is its wide support and integration with other tools, such as the cert-manager for certificate management and handling of the external DNS, which we will see soon. Also, the Kubernetes maintainer claims that there is no plan to deprecate Ingress as it perfectly supports simple web traffic in an uncomplicated manner.

Gateway API – the new way

In the autumn of 2023, the Gateway API became generally available. Instead of a single resource, the gateway API consists of multiple resource types following a pattern that was already used in other critical integrations:

- `Gateway`: Cluster entry point for incoming traffic

- `GatewayClass`: Defines the gateway control type that will handle the gateway

- `*Route`: Implements the traffic routing from the gateway to the service:

  - `HTTPRoute`

  - `GRPCRoute`

  - `TLSRoute`

  - `TCPRoute`

  - `UDPRoute`

Comparing the following diagram with the Ingress approach, we can see the two steps for incoming traffic going through the gateway and being redirected by `*Route`.

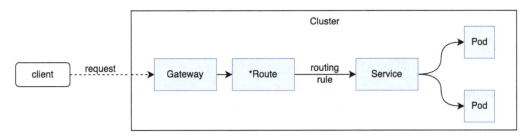

Figure 4.11: Gateway API

How do they compare?

	Ingress	Gateway API
Protocols	HTTP/HTTPS only	L4 and L7 protocols
Multi-Tenancy	Difficult/Custom Extensions	Multi-Tenant by design
Specifications	Annotation-based, controller-specific	Controller independent, own resource, standardized
Definition/Resource	Ingress resource	Gateway, GatewayClass, *Route resources
Routing	host/path-based	Supports header
Traffic Management	Limited to the vendor	Build-in/defined

Table 4.2: Comparing Ingress and the Gateway API

This rich set of features is a game-changer for those who have to build platforms. Previously, most of these capabilities had to be bought commercially or brutally forced into the cluster by some hacky workaround and self-developed tools. However, with the Gateway API, a wide range of protocols is supported, and it comes with two interesting additional features loaded.

You can create cross-namespace routes with the Gateway API, either with `ReferenceGrant` or with `AllowedRoutes`. The allowed routes are implemented via reference bindings, which need to be defined by the gateway owner as from which namespaces traffic is expected. In practice, the Gateway configuration will be extended as follows:

```
namespaces:
  from: Selector
  selector:
    matchExpressions:
    - key: kubernetes.io/metadata.name
      operator: In
      values:
```

```
    - alpha
    - omega
```

This will allow traffic from the `alpha` and `omega` namespaces.

Alternatively, we can use a `ReferenceGrant`, which is described as follows:

> *ReferenceGrant can be used to enable cross namespace references within Gateway API. In particular, Routes may forward traffic to backends in other namespaces, or Gateways may refer to Secrets in another namespace.*

They sound similar, but they aren't. Let's compare them:

- `ReferenceGrant`: A Gateway and Route in namespace A grant a service in namespace B to forward traffic

- `AllowedRoutes`: A Route to namespace B is configured in a Gateway in namespace C

Besides those cross-namespace additional security layers, the Gateway API also comes with its own extension capabilities. With them, you can define your own PolicyAttachment, such as BackendTLSPolicy, to validate the proper usage of TLS.

One last thing before we move on. The Gateway API comes with personas. Personas are pre-defined roles that allow fine-grained usage of gateway capabilities. Ingress just has one persona, a user, independent of whether it is an admin or a developer. The following table shows the write permissions of those personas in a four-tier model:

	GatewayClass	Gateway	Route
Infrastructure Provider	Yes	Yes	Yes
Cluster Operators	Sometimes	Yes	Yes
Application Admins	No	In Specified Namespaces	In Specified Namespaces
Application Developers	No	No	In Specified Namespaces

Figure 4.12: Write permissions for an advanced four-tier model

At this point, the Gateway API is the true savior of countless nights of getting inbound traffic right and building a multi-tenant platform with a clear and transparent approach.

ExternalDNS

The **ExternalDNS** project, developed by the Kubernetes contributors, is a tool that is often used within cloud-provided Kubernetes clusters, but still not often highlighted as a relevant implementation. Yet it bridges the gap between some random IPs of Pods in the cluster, takes it toward a proper DNS that is publicly or privately reachable, and routes traffic to the application within the platform. ExternalDNS provides support for almost every cloud and cloud-like environment, as well as traffic- and content-focused services such as CloudFlare.

However, ExternalDNS is not a DNS. Surprise! It reads the resources from the Kubernetes API and configures external DNS to point to the cluster's public endpoints. You could also say that ExternalDNS allows you to control DNS records dynamically via Kubernetes resources in a DNS provider-agnostic way.

Let's have a look at how ExternalDNS works together with the CoreDNS of Kubernetes and the Cloud DNS Service. In the next diagram, you can see the managed Kubernetes on the left. In this case, AWS EKS and the CoreDNS are running on Kubernetes to resolve internal DNS calls. When ExternalDNS is deployed, it observes the gateway, ingress, and service resources. When changes apply or new services come up, ExternalDNS updates the DNS records on the cloud provider or DNS service provider.

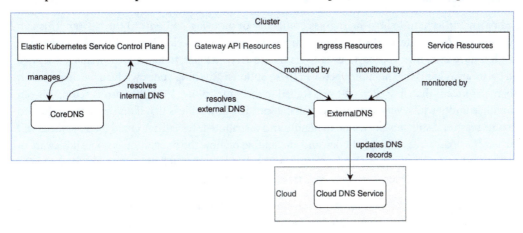

Figure 4.13: ExternalDNS

However, when looking for an alternative approach, you usually have only two fallback options: create the DNS entries by yourself (manually) or have it automated in some way with a custom controller or function, or during the infrastructure creation process.

Therefore, it is double painful to see the limitations of ExternalDNS:

- Missing fine-grained control; for example, ExternalDNS will create DNS records for all services and ingresses from any namespace

- ExternalDNS gives you only A records; if you need a TXT or CNAME record, you have to do it manually

- You will get default DNS configurations for the DNS name; otherwise, you have to manually define them

Besides that, you can also find increased costs (due to its outrageous DNS record creation behavior) and added complexity or latency. I can't fully agree with those types of issues as it is a question of how you handle them.

Consider using ExternalDNS only when you have mastered other parts of Kubernetes networking and are able to do fine-grained management of network policies and the gateway API so that you have strict control over which services are reachable, and how. In addition, consider enabling the DNSSEC feature and also establishing DNS monitoring.

Load balancing, EndpointSlices, and Topology-Aware Routing

Lastly, to round off the network segment, we will briefly discuss **load balancing**, **EndpointSlices**, and **Topology-Aware Routing**.

Load balancing is handled by the ingress controller or gateway API within the cluster. This can be outsourced to an external load balancer in combination with a cloud provider. All major cloud providers have their own approach, usually via their own controller, to manage the managed service load balancer. What is the difference between these options? Running your own load balancer within Kubernetes means that, first, all traffic gets routed to that entry point. With the correct setup, this can be multiple nodes with a very simple load distribution. The downside is that if one node is overloaded for some reason, it still gets the traffic to handle and distribute internally. A cloud load balancer will distribute the load across multiple nodes, and, depending on how the integration is done, it is aware of whether a node can handle more load or whether it should be redirected to another one. A downside of the public cloud load balancer is that you must also pay for it.

EndpointSlices become relevant for large-scale clusters. Endpoints are API objects that define a list of IP addresses and ports. These addresses belong to the Pods that are dynamically assigned to a service. When a service is created, Kubernetes automatically creates an associated Endpoint object. The Endpoint object maintains the Pods' IP addresses and port numbers that match the service's selector criteria. EndpointSlices were introduced in Kubernetes 1.16. They provide a way to distribute the network endpoints across multiple resources, reducing the load on the Kubernetes API server and improving the performance of large clusters. While in the past, a single large Endpoint object for a service became slow, multiple small EndpointSlice objects are now created, each representing a portion of the endpoints.

The control plane creates and manages EndpointSlices to have no more than 100 endpoints per slice. This can be changed, up to a maximum of 1,000. For the kube-proxy, an EndpointSlice is the source of truth for routing internal traffic.

EndpointSlices are also required for Topology-Aware Routing. With Topology-Aware Routing, you can keep the traffic within the cloud provider's **Availability Zone** (**AZ**). This has two main advantages: it reduces the network costs and improves the performance. Instead of a Pod in AZ 1 communicating with another Pod in AZ 2 and sending a lot of data, the Pod in AZ 1 will now talk to a replica (if available) that is also in AZ 1. To make this work best, the incoming traffic should be distributed evenly via an external load balancer and you should have at least three endpoints per zone. Otherwise, the controller will fail to assign that endpoint with a chance of around 50%.

Kubernetes as part of the platform control plane

Looking back to the second chapter, we can see in the reference architecture that a platform can be highly distributed, based on many services that, from a higher viewpoint, might not even belong together. We discussed that Kubernetes might often become a central part of your platform. However, how central can it become? As said, Kubernetes is not just there to run workload; it is a platform (based on promise theory and a standardized model and API) for building platforms. This shifts Kubernetes with one foot into the platform control plane and does this in two ways: as a resource controller and as a platform orchestrator.

Steering resources from within Kubernetes

The open source Crossplane project is the only provider-independent solution for managing cloud resources from within Kubernetes. Initially created to manage other Kubernetes clusters from within Kubernetes, it quickly became the universal solution for handling cloud resources. Cloud resources are available as CRDs and can be defined as Kubernetes-native resources through a specification file. This gives users the option to define what they need and leave the resource creation on the promise theory of Kubernetes. For the different clouds, so-called providers are available, which define the available resources. A user can create single resources or whole compositions, which are multiple resources that belong together.

The problem of external versus internally defined resources

What is the right approach to managing resources? Should they be defined by an infrastructure team or through input from demand forms? Can they be defined by the user via the platform? Anything is possible, but no simple answer exists.

Let's have a look at the two approaches. First, let us take a look at a scenario at Financial One ACME, coming from a more traditional, conservative background: infrastructure teams have been fighting over the last few years for automation and a declarative approach. While they manage their on-premises environments through Ansible, they decided to use something simpler for the cloud providers: Terraform (or OpenTofu). We will not go through the whole stack, but until we hit the Kubernetes platform, everything is orchestrated through classic CI/CD push principles and IaC via Terraform.

Financial One ACME is starting a new project to develop custom software for their internal usage. The team will utilize the ACME platform and work on the system's base architecture. They have defined that they will require certain file storage, a cache, a relational database, a notification service, and a message streaming service. As the platform provides some self-service, the team can copy the Terraform modules into their repository. From here, a predefined CI/CD pipeline will take over the configuration and deploy the resources in the defined environments. These self-defined but still externally managed resources are, in some ways, isolated from the rest of the system. They may live in the same repository or hierarchy and are managed by the team, but they are not strongly integrated.

On the other hand, they provide a certain level of stability. From within the platform user space, those resources are invisible, except that a discovery service exists. When an organization matures, the pipelines might be customized without notifying the owner or user. However, infrastructure and application are clearly separated, which is an advantage because of the totally different life cycles of the elements.

Fast forward and Financial One ACME has undergone a cloud-native transformation, leveraging platform engineering and IDPs to the maximum. Again, they plan to do a new internal project, which, of course, is completely different from the one before but somehow has exactly the same requirements. Some organizations' behavior will never change. This time, the project team created a new project in their developer portal. Automatically, all base requirements will be pushed into a new Git repository. The chosen resources are selected and pushed to the platform, where a controller decides where it deploys those resources. Some end up as managed services on the cloud provider, others in a shared service account from a specialized team, and a few in the project's namespace. After some time, the team understood that they had chosen the wrong configuration and, due to the adjustments, a migration to the new service took place. This scenario can be as true as the previous one but has different impacts.

The team needs to know in more detail which requirements they have; on the other hand, they have to trust in the predefined deployments and configurations. The platform engineering team has, in collaboration with the operational team, defined best practices with guardrails, ensuring operability but also matching almost all the requirements of the users. Within the cluster user space, the team can find all deployed resources and address them as a service within the platform, even though they are not running in it. However, sudden changes on the user side might cause resources to suddenly spike or get deleted in other places. Managed service teams have to handle such changes without warning. In total, all depending resources are acting extremely volatile and dynamic, hard to predict, and difficult to manage. On the other hand, the project can focus fully on the flow and progress as external resources are handled from within the cluster, rather than learning how to manage those with outer-cluster resource management solutions such as IaC.

Both approaches are fine. Both have pros and cons, and the one that is best for your organization often depends more on the human factor than any technological factor. However, what is clear is that internally defined cluster resources are more dynamic and shifted to the left, into the user's responsibility, than in externally defined resources.

In the end, it becomes a philosophical discussion. Externally defined resources are more traditional, whereas the internally defined approach is progressive and future-oriented. However, we don't have too many options to run cluster internal provisioning processes. Besides Crossplane, we have seen many meta-implementation controllers that read custom resources from the cluster and trigger CI/CD pipelines, for example. That's a poor workaround, if someone asks.

In this section, we looked at the fundamental capabilities that are required for Kubernetes, the challenges around them, and how we can solve them. Also, they don't feel like any of those crazy implementations you see at conferences. Getting these basics right will make the difference and decide whether everything else on top will be a pleasure or a pain. Up next, we will close this chapter by looking into the approach of finding the right node sizes and the ramifications for flexibility and reliability.

Designing for flexibility, reliability, and robustness

In the previous part, we discussed cluster scalability and how the VPA, HPA, and CA play together. This helps to create a flexible, reliable, and robust system. A key part of this is also allowing customization as long as it doesn't harm your system. Components of the cluster must play together seamlessly but must also be exchangeable where needed. This is sensitive: you have, on the one hand, a breathing cluster that grows and shrinks its demand over time; then, you have all the extensions on and around the cluster that allow you to serve the best possible feature set for your use case. You also have the continuously evolving open source community that frequently delivers updates and new developments, which should be integrated and made available for your users. As we told you earlier, this is why you must have a product mindset – to build the best possible platform for your users. Throw away things you don't need or that are outdated, but keep the whole system as your core.

Now, for some reason, we still see discussions about whether you should put all your workload on Kubernetes, and whether it is reliable. We have to look at this discussion from two perspectives: bottom-up from Kubernetes's infrastructure and core responsibilities, and top-down from what the user can see and experience.

Optimize consumption versus leaving enough head space

As we learned in *Chapter 2*, the Kubernetes cluster and the workload it manages have an interesting relationship and influence on each other. Some applications require more CPU power, others scale up instead, and the rest just run on demand. Finding the right match isn't easy, but an ideal target is high resource consumption as it optimizes the usage of the available resources and therefore the costs.

How to make the clusters the right size

Evaluating the right size for the cluster is always a challenge. Finding the right solution is a clear case of *it depends*. Let's take Financial One ACME, which has to provide a new cluster. We don't know a lot about the expected workload, just that it requires some memory and has a few moving parts that are not that resource-demanding. So, we can take one of the following options:

- 1x 16 vCPU, 64 GB memory
- 2x 8 vCPU, 32 GB memory
- 4x 4vCPU, 16 GB memory
- 8x 2vCPU, 8 GB memory

For availability reasons, the first option would be a bad idea. If you run an update in the cluster and the whole system goes down or you have to provision one new node, you need to shift all the apps over and shut down the old one. Option 4 comes with many nodes. Due to its small size, this can lead to a resource shortage as some base components that are required for the cluster will consume a part of the CPU and memory. In addition, depending on the cloud provider, it might be that you have other limitations, such as available IP addresses, bandwidth, and storage capacity. Also, if you have an app that needs 1 vCPU, and might scale to 1.5 – 2.0 vCPU, it would kill an entire node.

However, how many resources are used per node by Kubernetes? By default, for the CPU, we can use the following rules:

- 6% of the first core
- 1% of the second core
- 0.5% of the next two cores
- 0.25% from the 5th core onward

We also have some rough rules for the memory:

- 25% of the first 4 GB
- 20% of the following 4 GB
- 10% of the next 8 GB
- 6% of the next 112 GB (up to 128 GB)
- 2% of anything above 128 GB
- In case the node has less than 1GB of memory, it is 255 MiB

In addition, every node has an eviction threshold of 100 MB. If the resources are completely utilized and the threshold is crossed, Kubernetes starts cleaning up some Pods to prevent completely running out of memory. At learnk8s, you can find a very detailed blog about it (`https://learnk8s.io/kubernetes-node-size`).

Let's visualize these numbers:

	2 vCPU 8 GB RAM	4 vCPU 16 GB RAM	8 vCPU 32 GB RAM
Kubelet + OS vCPU	70 m or 0.07 vCPU	80 m or 0.08 vCPU	90 m or 0.09 vCPU
Kubelet + OS memory	1.8 GB	2.6 GB	3.56 GB
Eviction threshold	100 MB	100 MB	100 MB
Available vCPU	1930 m or 1.9 vCPU	3920 m or 3.9 vCPU	7910n or 7.9 vCPU
Available memory	6.1 GB	13.3 GB	28.34 GB

Table 4.3: Resource consumption and available resources for different node sizes

From those available resources, you also have to take away anything that is required to cluster, such as DaemonSet for logging and monitoring, kube-proxy, storage driver, and so on.

Doing the math, if you have many small nodes, the basic layer of used resources is larger than that of the one with big nodes. Let us look at an example. Two 2vCPU 8 GB RAM nodes require 140 m of CPU and 3.6 GB of RAM in total, while one 4vCPU 16 GB RAM only requires 80 m of CPU and 2.6 GB of RAM. Those are the costs for higher availability, but in a side-by-side comparison of just the available resources, a single node requires fewer resources to be assigned to Kubernetes. However, ideally, a node is utilized at its maximum capacity; we don't run a virtual machine where 80% head space of non-utilized resources is standard! At least 80% utilization of a node would be the target to get the best performance/price ratio. This is also because of the energy proportionality of a server. The energy needed by a CPU doesn't linearly scale with the workload.

Imagine that a server can consume up to 200 watts. Then, most servers would need between 150 and 180 watts for around 50% of CPU utilization. This means the more a server is utilized, the better the energy/CPU utilization ratio, which is also more cost-efficient and sustainable.

We have to consider other factors while choosing the node sizes for the cluster:

- If you provide just a few large nodes, but the user defines an anti-affinity so that on each node only one Pod of a replica is running and you don't have enough nodes because the expected replicas are too high, then some Pods become unschedulable.

- Large nodes tend to be underutilized, so you spend more money on something you don't use.

- If you have too many small nodes and you continuously run into pending Pods for which each cluster has to scale, that might make the users unhappy as it always takes time to scale up new nodes. Also, it might affect the servability of an app if it runs continuously under resource limitations.

- Many nodes cause a higher amount of network communication, container image pulls, and duplicate image storage on each node. Otherwise, in the worst case, you always pull an image from a registry again. With a small cluster, that isn't a problem, but with a large number of nodes, this becomes quite chatty.

- The more nodes there are, the more communication happens between them and the control plane. At some level of requests, this means increasing the node sizes of the control plane.

Finding the right node and cluster size is a science in itself. Neither end of the extreme is good. Start looking at the kind of workload you expect. If you're not sure, start with something medium-sized and adjust the node sizes if needed. Also, consider always having a little bit more memory available. The core components of Kubernetes per node don't require a lot of CPU, but do require at least something between 2 and 3 GB of memory plus all the other default components.

Solution space

Most public cloud providers come with the ability to run multiple instance types for Kubernetes nodes. This is helpful for different use cases, from isolating workloads to optimizing the utilization or migrating from one CPU architecture to another. We also talked about GPU utilization, which you can combine in such scenarios to run non-GPU workloads on CPU nodes while doing model training on the GPU instances.

To do this, you have to manage and label the nodes correctly, and provide users with a transparent approach and support for defining their affinity settings.

To identify the right node sizes, learnk8s also provides you with an instance calculator (`https://learnk8s.io/kubernetes-instance-calculator`), which you might consider before you start building your own Excel sheet for doing the math.

Also, when defining the cluster size and scale doesn't look that relevant, with the right node sizes, you can have a direct impact on costs, utilization, application availability, user experience, and how many additional implementations you have to do to compensate for potential drawbacks.

Summary

In this chapter, we got closer to some of the relevant components of Kubernetes as the cornerstone for your platform. We first examined whether Kubernetes is the right choice, and also looked at why it often is the right way to go. With the promise theory at its heart and many robust features for running and extending a platform, Kubernetes is an almost perfect foundation for a platform. From here, we looked into some of the very basic elements of Kubernetes: storage, networking, CPU architectures, and GPU support. In this context, we learned about some design considerations and problems we might face while implementing Kubernetes. While Kubernetes as a foundation might feel different in every environment, it is possible to create a unified experience. This will come with major drawbacks, such as losing the features of certain cloud providers, flexibility, and customizability.

Next, we discussed finding the balance between a very stiff and highly flexible system. Both can be seen as robust and reliable, but they come with very different challenges and problems. Therefore, we did a short thought experiment to find the right cluster sizes and node types before we closed this section by discussing approaches for implementing guardrails for the user space. This helps us provide flexibility within the user space but protects the platform from misbehavior and wrongly configured services by users. We learned about this in this chapter.

In the next chapter, we will focus on the automation of platforms. Besides the infrastructure, automation is a critical component of a platform and, as you will see later on, can be a bottleneck and cost driver in the long run. You will learn how to design a proper release process, how to implement it in CI/CD and GitOps, and how to use this combination for the life cycle of the platform artifacts. We will also show you how to effectively observe this process.

Further Reading

- [1] Objects in Kubernetes: `https://kubernetes.io/docs/concepts/overview/working-with-objects/`

- [2] Karpenter: `https://karpenter.sh/`

- [3] CSI drivers index: `https://kubernetes-csi.github.io/docs/drivers.html`

- [4] RISC-V Kubernetes nodes on Scaleway: `https://www.scaleway.com/en/docs/bare-metal/elastic-metal/how-to/kubernetes-on-riscv/`

- [5] SpinKube: `https://www.spinkube.dev/`

- [6] SpinKube Overview: `https://www.spinkube.dev/docs/overview`

- [7] Nvidia – Improving GPU Utilization in Kubernetes: `https://developer.nvidia.com/blog/improving-gpu-utilization-in-kubernetes/`

- [8] DRA with GPU:

  - `https://docs.google.com/document/d/1BNWqgx_SmZDi-va_V31v3DnuVwYnF2EmN7D-O_fB6Oo/edit#heading=h.bxuci8gx6hna`

  - `https://static.sched.com/hosted_files/colocatedeventseu2024/83/Best%20Practices%20for%20LLM%20Serving%20with%20DRA.pdf`

  - `https://github.com/NVIDIA/k8s-dra-driver?tab=re`

Get This Book's PDF Version and Exclusive Extras

Scan the QR code (or go to `packtpub.com/unlock`). Search for this book by name, confirm the edition, and then follow the steps on the page.

Note: Keep your invoice handy. Purchases made directly from Packt don't require one.

5

Integration, Delivery, and Deployment – Automation is Ubiquitous

Chapter 2 included a reference architecture of a platform highlighting layers such as *Developer Experience*, *Automation and Orchestration*, and *Observability Plane*. *Chapter 3* ended with a different perspective on this reference architecture using a top-down approach from *Purpose, User Interface, Core Platform Components, Platform as a Product,* and *Success KPIs*.

Most platforms are built with the purpose of making it easier for development teams to ship software without having to deal with all the complexity around building, deploying, testing, validating, securing, operating, releasing, or scaling software. In this chapter, we will dive into those layers and components of our platform so we understand how we can centralize and automate the expertise it takes to ship software and provide it as a self-service.

By the end of this chapter, we will have learned how to define an end-to-end release process for software artifacts, align the **continuous integration / continuous deployment (CI/CD)** process with the artifact life cycle phases, automate the phases using tools for CI, CD, and continuous release, integrate those tools into your existing process, observe the automation, and scale through an IDP.

As such, we will cover the following main topics in the chapter:

- An introduction to Continuous X
- GitOps: Moving from pushing to pulling the desired state
- Understanding the importance of container and artifact registries as entry points
- Defining the release process and management
- Achieving sustainable CI/CD for DevOps – application life cycle orchestration
- **Internal Developer Platforms (IDPs)** – the automation Kraken in the platform

An introduction to Continuous X

If this is the first time you are hearing about **Continuous Integration (CI)** or **Continuous Delivery (CD)**, then we suggest checking out some of the great literature that exists on these basic concepts. *Jez Humble* is the maintainer of `https://continuousdelivery.com/` and has co-authored the original *Continuous Delivery* book with *Dave Farley*. If you need a crash course on this topic, then please have a look at their material. There are also several recorded talks that give a great overview, such as the one titled *Continuous Delivery Sounds Great But It Won't Work Here*: `https://vimeo.com/193849732`.

So that we all have the same common understanding, let us quickly recap what the building blocks are and why this is important in software delivery.

High-level definition of Continuous X

The basic foundation of automating software delivery is to have proper *configuration management* of all assets required to build, deploy, validate, operate, and scale our systems: code, tests, deployment and infrastructure definitions, dependencies, observability, ownership, and more. Putting all those assets into version control allows us to generate repeatable and reliable output, gives us auditability, allows us to revert breaking changes, and gives us disaster recovery capabilities.

Git, or any type of Git flavor, is most likely what can be found today in software organizations for version control. Depending on the Git solution you use, you will see additional built-in capabilities such as cross-team collaboration (issue tracking, resolving merge conflicts, etc.), automation (commit checks, delivery pipelines, etc.), or reporting (efficiency or DORA metrics).

Now that we have established the foundation, let's dive into Continuous X.

CI

CI is the practice that emphasizes frequent and automated integration of code changes into a shared repository. When multiple developers work on the same code base, it's important to bring those changes together frequently to validate that the code integrates well and produces an artifact (a container image, a binary) that can be deployed into an environment. Key aspects of CI include the following:

- **Automated builds**: Code commits in version control trigger an automated build process that compiles, runs tests, and generates artifacts
- **Test automation**: Unit tests, integration tests, and other checks are executed during the build process marking the build broken if any tests should fail
- **Feedback loop**: This provides rapid feedback to developers to quickly fix issues, leading to overall higher code quality and stability

Continuous testing and validation of observability

While CI already highlights the importance of automated unit and integration tests, we want to stress the fact that more automated testing and validation early in the life cycle will result in better quality and stability. Assuming the built software provides REST APIs or a UI, basic validation against those interfaces should be done to validate the accurate functionality (e.g, do APIs return proper responses, are the correct HTTP status codes used when testing with invalidated parameters, are there any timeouts in API calls, or are any HTTP errors returned?).

Observability is essential to validate if systems are healthy and provides the data to troubleshoot problems faster. As part of the continuous validation process, we must validate that the tested build is producing valid and expected observability data. We should validate that all expected metrics, logs, or traces are produced and that there is no obvious anomaly or outlier present after running those basic unit, integration, or API tests.

We have been stressing that observability has to be a non-functional requirement of modern software. This is why the CI should already validate if the expected data is produced. If not, then this is just like a failing unit or integration test and you should mark the build as broken!

For Financial One ACME and its critical financial services, we should validate the following:

- Will the API properly validate the access control of the caller (e.g., not be able to query financial data from other users)?

- Will the API not log any confidential data, such as credit card numbers, usernames, or tokens?

- Will the API properly generate metrics for failed attempts so this can be used in production to alert on potential hack attacks?

Continuous delivery

As defined on the continuous delivery site (`https://continuousdelivery.com/`), "*Continuous Delivery is the ability to get changes of all types—including new features, configuration changes, bug fixes and experiments—into production, or into the hands of users, safely and quickly in a sustainable way.*" The goal is to take the fear of failure out of deployments. Instead of infrequent big bang releases, deployments should become routine as they continuously happen. Additionally, applying new deployment patterns such as blue/green, canary, or feature flagging allows us to reduce the risk even further. These will be discussed further in the section about deployments versus releases!

Some aspects of continuous delivery are as follows:

- **Automated deployments**: New artifacts that come out of CI are bundled with other changes and get automatically deployed. Deployment definitions are declarative and version-controlled and therefore allow a more predictable, repeatable, and low-risk way of updating in any environment.

- **Deployment pipelines**: Pipelines allow higher-level testing and deployment validation as compared to CI. Here is where performance, security, scalability, resiliency, and user experience tests get executed. This validates not just a single artifact but the full deployment change set!

- **Quality gates and promotion**: At the end of a deployment pipeline, all test results act as a quality gate before promoting that change into the next environment: from development to **Quality Assurance (QA,)** from QA to staging, from staging to production.

- **Rolling back versus rolling forward**: If the quality gate fails in production, a rollback can be triggered by reverting back to the previous version-controlled deployment configuration. Another strategy is rolling forward, which means that problems are fixed, and thanks to automated deployments, the fix can be deployed quickly to avoid the need for a rollback.

Continuous deployments – decoupling deployments from releases

CD deploys changes in an automated and fast way. However, there is still a risk that a change results in a failure, requiring either rolling backward or rolling forward, as explained earlier.

Continuous deployments go a step further and embrace new deployment patterns that favor the separation of the deployment of a change and releasing the new feature set to the end users. The current well-established patterns are as follows:

- **Blue/green deployments**: The new deployment (commonly labeled *blue*) will be done in parallel to the existing deployment (commonly labeled *green*). Through a load balancer, traffic can be switched to blue. If there is a problem with blue, traffic can be switched back to the still-running instance, therefore eliminating the need for a rollback while minimizing the impact on the end user. If all goes well, green becomes blue until the next deployment comes along.

- **Canary deployments**: Similar to blue/green deployments but on a more granular level. It's the practice of a staged deployment of the new version besides the old version. First, it is deployed to a small subset of users or a percentage of the traffic. If everything is good, the staged rollout continues until all user traffic has the new version. If a problem occurs during the stages, the old version will receive the traffic.

- **Feature flagging**: Instead of load-balanced side-by-side deployments of the old and new versions, feature flagging allows developers to "hide" new code behind a switch/toggle. During a deployment, the new version gets deployed over the old one without executing the new hidden code. Through fine-grained configuration, features can be turned on for individual users, user types, geographical regions, or any other attribute of a consumer of a service. If a feature has a problem, it only takes a runtime configuration change and that code becomes inactive again.

Decoupling deployments from releases allows teams to better control the rollout of new features of their software and with that, minimize risk. There is more to explain about implementation details as well as challenges but that's beyond the scope of this book. If you are interested in learning more, look into *OpenFeature [1]*, a **Cloud Native Computing Foundation** (**CNCF**) incubating project. OpenFeature provides a standard, feature-flag management, vendor-agnostic way for developers to implement feature flags. The community around it also has a lot of best practices around progressive delivery, which includes all the patterns discussed previously.

Continuous X for infrastructure

Continuous X is not only relevant for application code or configuration. The same concepts should be applied to any infrastructure definition. As a platform, we will need certain infrastructure components. Whether that's virtual machines, Kubernetes nodes, load balancers, **Domain Name System** (**DNS**), file storage, databases, virtual networks, serverless, or any other component that allows us to run our core platform as well as the applications that will be deployed, operated, and managed through our platform self-services.

Just like with application code, we want to configure our infrastructure requirements as code, version control them, and apply the same CI and continuous testing, validation, and delivery.

GitOps is also a term that emerged over the past years and it focuses on automating the process of provisioning infrastructure from a desired state defined declaratively and version-controlled in Git. We will cover GitOps in more detail in a later section of this chapter. First, let's discuss the basics by starting with **infrastructure as code** (**IaC**).

IaC

There are many different tools that enable **IaC** and most likely, you already have one or several in your organization: Terraform, Ansible, Puppet, Chef, CloudFormation, and **AWS Cloud Development Kit** (**CDK**) just to name a few.

Here is a very simple Terraform snippet that would create an EC2 instance of type c5.large:

```
resource "aws_instance" "finoneacme_demoserver" {
  ami             = "ami-01e815680a0bbe597"
  instance_type = "c5.large"
  tags = {
    Name = "IaCTFExample"
  }
}
```

Here is one more example of creating an AWS S3 bucket using Ansible:

```
- name: provisioning S3 Bucket using Ansible playbook
  hosts: localhost
  connection: local
  gather_facts: false
  tags: provisioning
  tasks:
    - name: create S3 bucket
      S3_bucket:
        name: finoneacme_bucket_dev
        region: us-east-1
        versioning: yes
        tags:
          name: bucketenv
          type: dev
```

IaC enables us to define the desired state of our IaC. This is code that can be version-controlled, such as application code, and once deployed, it results in the desired infrastructure being provisioned. Like with application code, we can use the following:

- **CI**: Use this to validate that all our IaC is valid. IaC tools typically have features to "dry run" and validate that there is no mistake and that all config files have no conflict or dependency issues.

- **Testing and deployment validation**: After IaC is deployed, we can validate that we really got our desired state (e.g., ensuring that the EC2 instance is really up and running, that the S3 buckets are accessible, etc.).

- **Rollback or revert**: IaC gives us the option to roll back changes or revert to a previous version because everything is version-controlled!

For more details on IaC, including version control strategies (where IaC lives), the authors recommend looking into existing books and blogs on that topic.

Crossplane – IaC for platform and applications

IaC is not limited to the core platform services but is also relevant for the applications that we allow our development teams to deploy through our platform self-services. A new application may need file storage, a database, and a public DNS, or need to deploy a third-party solution; it depends on its own virtual machine, which is accessible from the deployed app that may reside on K8s.

You can provide templates for Terraform, Ansible, and CDK, which your developers can add to their own code repositories, and which are then applied as part of the Continuous X of their application.

One tool that has emerged in the cloud native space to cover both application and infrastructure orchestration is the **CNCF project Crossplane** *[2]*. Besides coming with a lot of different providers for all major cloud vendors or even Terraform, it comes with Compositions. **Compositions** are a template for creating multiple managed resources as a single object. This allows the platform team to define such templates for common application architectures and then have the application team simply use that template to instantiate the correct infrastructure and deploy the application.

One of our self-service use cases discussed in *Chapter 2* was the automated provisioning of a performance test environment. We could define a composition that would be as easy to use by the development teams as the one shown in this example:

```
apiVersion: composites.financialone.acme/v1alpha1
kind: PerformanceTestCluster
metadata:
  name: ptest-devteam1
spec:
  clusterSize: "medium"
  targetApp:
    repoUrl: "https://financialone.acme/our-app-repo"
    targetRevision: "2.5.1"
    chart: "ourapp"
  loadProfile: "spike-load"
  observability: true
  notifySlackOnReady: "#devteam1"
  leaseTime: "12h"
```

The Composition definition of `PerformanceTestCluster` would have been created by the platform engineering team in combination with those experts who know how to install the load testing and observability tools. In the preceding example, a new medium-sized K8s cluster would be provisioned, the requested app would be installed using the referenced Helm chart, observability data would be captured (e.g., Prometheus and log scraping configured) and the load testing tool would be deployed to be able to run spike load scenarios. Once everything is ready, a Slack notification with the environment details will be sent to the team. Last but not least, the environment would also be shut down after 12 hours as specified in the mandatory `leaseTime` field.

The preceding example already shows the power of IaC when integrating this into our Continuous X efforts!

Continuous X as a system-critical component in our platform

There are many different tools we can choose from to implement Continuous X: Jenkins, GitLab, Tekton, Argo CD, Flux, Keptn, Crossplane, Selenium, and k6, to just name a few. Whatever tools we choose, those tools need to be available, resilient, and secure all of the time, as they are the backbone of our platform. Those tools are as business-critical as any software that powers the business we are in. Think about Financial One ACME. If the developers need to push out a fix to a critical production issue on their financial software, they need Continuous X to work perfectly.

To ensure that those components are available when they are needed, we need to apply the same best practices as we put on our business-critical apps:

- **Secure by default**: If attackers find their way into our Continuous X toolkit, they have open doors to enter any system that is managed by our platform. Because of this criticality, *Chapter 7* is fully dedicated to building secure and compliant products.

- **Test every change we make**: Let's assume we use GitLab as one of our tools for Git and CI. We must version control the deployment configuration of GitLab and run it through the same Continuous X process to validate every new version or configuration change. If necessary, we will roll back or roll forward in case an update is causing issues!

- **Deploy highly available**: Follow the deployment guidelines for those tools for high availability. If we have globally distributed teams, we want to make sure to deploy certain components as close as possible to our end users. Also, look into zero-downtime upgrade options and follow them.

- **Observe each component**: Every tool provides some type of telemetry data that indicates health. Argo CD, for instance, exposes Prometheus metrics for work queue length, Git requests, and sync activity. Those give an indication of whether Argo CD is still able to do its job. A constantly increasing work queue depth is a sign that Argo CD can't keep up with all requests, which needs to be looked into.

- **Service-level agreements (SLAs) and alerting on problems**: We – the platform team – must know that something is wrong before our end users report it to us. That's why we need to set up SLAs for each component and configure proper alerting in case systems are not working as expected. The simplest way to do this is to set up synthetic checks against the key API endpoints of each tool (e.g., validate that Jenkins UI is responsive with a synthetic check that runs every five minutes; this gives us an early warning signal in case Jenkins starts having problems before anyone else notices it).

The following is an example dashboard highlighting key health indicators of tools, such as Argo CD. The same must be applied to all other tools that make up our core platform capabilities!

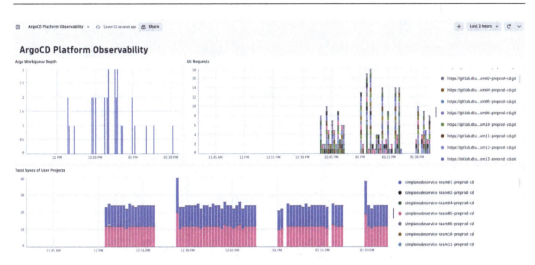

Figure 5.1: Monitor and observe every tool part of the platform

> **Platform Components are as business-critical as your critical apps**
>
> The core of the platform will revolve around deploying changes into various environments. All tools that support those use cases – especially those for Continuous X – need to be highly available, resilient, and secure. Make sure to apply the same engineering best practices to all components of the platform!

Now that we have recapped the core concepts of Continuous X (integration, testing, validation, delivery, deployment, and release) and discussed that all configurations (code, deployment config, infrastructure, observability, etc.) must follow the same principles, it's time to have a look into how these concepts can be used to provide self-service autonomy to teams that leverage this through a platform.

GitOps – Moving from pushing to pulling the desired state

CI/CD has been around for many years. The *Continuous Delivery* book was initially released back in 2010 – years before the emergence of containers (made popular through Docker, starting in 2013) and container orchestration platforms (such as Kubernetes, starting in 2014).

Fast forward to 2024 when this book was initially published; we live in a world where the following is the case:

- **Git** is the source of truth. It contains everything as code: source code, tests, infrastructure and deployment definitions, observability, ownership, and so on.

- **CI/CD** is building, testing, and packaging container / **Open Container Initiative** (**OCI**) images from a source code Git repository and publishing them to an artifact registry (e.g., Harbor).

- **GitOps** continuously attempts to apply the latest desired state as declared in a deployment Git repository (e.g., Helm Charts) on the target Kubernetes environment and pushes any additional configuration (e.g., observability to the respective tools).

The preceding description is not a one-size-fits-all model. GitOps will be implemented slightly differently in every organization. If you search for `What is GitOps?`, you will find many different variations, such as the following:

- **Separate CI and CD**: CI publishing containers and CD publishing packaged artifacts, such as Helm charts
- **Single Git**: Everything as code (source, test, deployment, observability, etc.) in a single Git repository
- **Push GitOps**: Pushing the desired state through pipelines or workflows versus pulling changes into the target environment through GitOps operators

In the following section, we will shine the light on one flavor of GitOps that favors the **pull** (*pulling a configuration into the target environment*) model over the **push** (*pushing a configuration from an external tool into the target environment*) model, which can be implemented with CNCF tools, such as Flux or Argo CD, as shown in the following illustration taken from Codefresh's Learning Center on GitOps: `https://codefresh.io/learn/gitops/`.

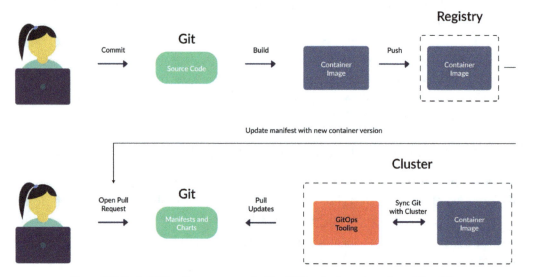

Figure 5.2: Basic GitOps flow as promoted by GitOps tools such as Argo CD or Flux

It's your choice as to whether you favor the Push model using automated pipelines or workflows to push changes into your target Kubernetes environments! To make the decision easier, let's dig into more detail on the individual phases and learn about some best practices that should happen in every phase!

Phase 1 – from source code to container image

Moving to GitOps doesn't change anything about how we are building our applications from source code that is version-controlled in Git. Before building new artifacts, a good practice is to automate dependency checks and updates. Tools, such as **Renovate Bot** *[3]*, integrate with Git and create Pull Requests in case outdated dependencies, third-party libraries, or other dependencies are found.

Once we have our up-to-date code in Git, **CI** has the same job as it always had: it creates an artifact, most likely a container image, that gets pushed to a container registry! That CI can be triggered on demand, on commits to certain branches, on Pull Requests, or on any other trigger that makes sense for the organization!

There are many different tools available that can do the job: Jenkins, GitLab, Azure DevOps, GitHub Actions, Bitbucket Pipelines, and many more. There are also various container registry options. We will talk about container registries and the importance of them a bit later in this book as they are – like all the components of our platform – critical to the success of our future platform!

Once the CI has successfully compiled the source code, executed unit tests, and run any additional quality checks on the code, it's time to package the binary into a container image. There are many best practices out there for building container images; from just packaging a single service into a container, being careful to use public base images, to building the smallest image possible. Following all those rules will lead to a higher rate of success as the container image moves from CI all the way into production. Here are two obvious but very important practices that we want to mention in this chapter (for more information, please refer to the books previously mentioned):

- **Properly and consistently tag your images**: Images are generally identified by two components: their name and their tag (e.g., `finoneacme/fund-transfer:2.34.3`, where `finoneacme/fund-transfer` is the name and `2.34.3` is the tag). When you build an image, it's up to you to tag it properly but you should follow a consistent policy. There are two common ways to do this that are used in the industry:

  - **Semantic versioning**: A common way is using a version number. The **Semantic Versioning Specification** *[4]* provides an easy-to-follow guideline with a three-part version number: `MAJOR.MINOR.PATCH`. Every minor or patch version number must be for a backward-compatible change. In combination with the version name `latest`, which is the default to point to the latest available version, you can provide easy access to a specific image (e.g., `X.Y.Z`). It also allows for the selection of the latest patch release for a specific minor release (e.g., `X.Y`) or the latest minor and patch version for a major release (e.g., `X`).

  - **Tagging using Git commit hash**: When Git commits are triggering a CI build, it is best to directly use the `git commit` hash as version instead of keeping track of semantic versioning. The commit hash on the image also immediately tells us which source code commit was responsible for producing this container image. For our image, this could mean it's tagged like this: `financeoneacme/fund-transfer:d85aaef`.

- **Scan your images for known vulnerabilities**: Images that are built by your CI must be scanned for known vulnerabilities. This can be done as part of an extra step in CI (e.g., call a scanning tool before uploading the image to the registry). It can also be handled by the container registry itself. Depending on the registry used, the image can be scanned during upload and blocked or quarantined if vulnerabilities are found. This stops known vulnerabilities at the time of creating the container image. Be aware that this does not stop vulnerabilities that are identified at a later time, such as the well-known `log4j` vulnerability. This is why security must go beyond the static checks in the CI process and continue throughout the life cycle of an artifact. This will be explored more in *Chapter 7*, which dives deeper into building and operating secure products.

Now that we have a container image uploaded to our container registry, we can use it in a deployment definition.

Phase 2 – from container image to metadata-enriched deployment artifact

The container image now needs to find its way into a deployment definition. Looking at Kubernetes, we would need a manifest file that defines our deployment definition, which we can later apply to our K8s cluster.

In *Chapter 3*, we highlighted that adding metadata (ownership, observability level, notifications) to our deployment file will benefit lots of the self-service use cases, such as the *Requesting Logs as Self-Service* use case. Besides our source code, we therefore also need to version control our deployment definition and should enforce a minimum set of metadata. This enables our platform self-service use cases (e.g., who has deployed which services belong to which application) as well as following general infrastructure best practices (e.g., defining request and resource limits), as you can see in the following manifest:

mainfest.yaml (Kubernetes Deployment)

```yaml
apiVersion: apps/v1
kind: Deployment
metadata:
  name: fund-transfer-service
spec:
  ...
  template:
    metadata:
      annotations:
        # team ownership
```

```
      owner.team: dev-team-backend
      app.kubernetes.io/name: fund-transfer-service
      # app-context
      app.kubernetes.io/part-of: backend-services
      app.kubernetes.io/version: d85aaef
      # log level
      observability.logs.level: info
      # log source
      observability.logs.location: stdout
   ...
spec:
  containers:
  - name: fund-transfer
    image: "financeoneacme/fund-transfer:d85aaef"
    imagePullPolicy: IfNotPresent
    resources:
      # specify limits
      limits:
        memory: "200Mi"
        cpu: "2"
      requests:
        memory: "100Mi"
        cpu: "2"
   ...
```

The deployment definition can be a manifest or a set of manifest files, as we likely also need additional K8s resources, such as a service or Ingress definition. Instead of plain manifest files that we organize in a subfolder next to the source code of our service, there are options to leverage templating or packaging frameworks and tools, such as **Kustomize** *[5]* or **Helm** *[6]*.

It is important to consider how to organize and version control all of our everything as code files. There are several strategies with pros and cons that are well documented. To learn more, the authors suggest reading up on the different patterns for repositories and directory structures. The following section is just a quick overview of what you must know to continue your in-depth research on this topic!

Monorepo versus polyrepo

When designing your git repository structure, you have two general options which the industry refers to as monorepo or polyrepo; a single Git repository or multiple repositories split by teams or functions:

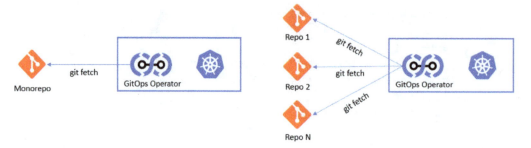

Figure 5.3: Mono versus polyrepo – pros and cons for both patterns

- **Monorepo**: In the monorepo pattern, all configuration files (code, infrastructure, observability, etc.) are stored in a single Git repository. This pattern applies to every potential environment (dev, testing, staging, production)—meaning every configuration is in a single Git repo, separated by folders.

  The benefit of a monorepo is that everything is in one place. The downside is that this repo will eventually become very large, resulting in potential performance issues when tools like ArgoCD or Flux need to constantly scan the entire repo for changes. It also makes it harder to separate the concerns between application and infrastructure owners.

- **Polyrepo**: In the polyrepo pattern, we have multiple repositories, which makes it easier to separate concerns. These can be repositories per team (app team A, B, C), per environment (dev, staging, prod, etc.), per tenant (in multi-tenancy systems), or per organizational boundary (app teams, platform teams, etc.).

  The benefit is a better separation of concerns. The downside is that we eventually end up managing a large number of Git repositories, which makes it harder to ensure consistency and overall validity across all configurations spread across multiple repos that make up the entire configuration.

Directory structure – follow the DRY principle

Whether mono or polyrepo, the individual Git repository will need to have a good directory structure. Typically, we see it reflecting the organizational structure separating the development teams from the teams that manage the underlying infrastructure or platform. A good practice is to follow the **don't repeat yourself** (**DRY**) [7] principle. The idea is to find the best structure to avoid copying/pasting common YAML settings in different places. This is where tools such as Helm and Kustomize help.

The following structure example was inspired by a GitOps blog from **Red Hat** *[8]* and shows the configuration for the platform administrator and the configuration for an application team using a templating tool such as Kustomize. This structure could either be used in a single monorepo separated by a folder, or it could be put into individual repos following the polyrepo pattern.

Platform Team	Development Team
├── bootstrap	└── deploy
│ ├── base	├── base
│ └── overlays	└── overlays
│ └── default	├── dev
├── cluster-config	├── prod
│ ├── gitops-controller	└── stage
│ ├── identity-provider	
│ └── image-scanner	
└── components	
├── applicationsets	
├── applications	
└── argocdproj	

Table 5.1: One possible directory structure used by tools such as Kustomize

Now that we have learned about the different ways to organize our repositories and directory structures, we need to learn how the new container image in *Phase 1* makes it to our deployment definition in our repositories!

Updating manifest with new container image version

The last piece to *Phase 2* is to promote a new image version that came out of CI into our deployment files. This could be updating the deployment manifest, as shown in the preceding example, or updating a `values.yaml` file when using Helm charts.

As those configuration files are version-controlled in a Git repository (mono or poly, it doesn't matter), we should follow the regular Git workflow and create a Pull Request to promote the update of that image. We can automate the creation of that Pull Request in two ways:

- **Step in CI pipeline**: As the last step in CI, a Pull Request can be opened, promoting the new image tag to the respective deployment definition repo and the right directory structure (for example, updating the values in the development overlay directory if this is a newly built image)

- **Webhook in a container registry**: When the registry receives a new image from CI, it can trigger a webhook, which allows us to create that same Pull Request in our deployment Git repository

The following shows the updated version information of the *Kubernetes Deployment* that we have seen in the earlier example. You can compare this with the values of the previous example:

```
apiVersion: apps/v1
kind: Deployment
metadata:
  name: fund-transfer-service
spec:
  ...
  template:
    metadata:
      annotations:
        owner.team: dev-team-backend
        app.kubernetes.io/name: fund-transfer-service
        app.kubernetes.io/part-of: backend-services
        app.kubernetes.io/version: a3456bc
        observability.logs.level: info
        observability.logs.location: stdout
    ...
    spec:
      containers:
      - name: fund-transfer
        image: "financeoneacme/fund-transfer:a3456bc"
    ...
```

Now, we know how a new image that is built by CI makes it into the deployment definition in our respective repositories. We already saw the GitOps operator in an image earlier in this chapter. Now, it's time to dive into what the GitOps operator really does to apply our desired state from Git to our target clusters!

Phase 3 – GitOps – keeping your desired deployment state

Now that we have everything as code in Git, it's time to talk about how to apply that configuration to our target environments. One way is to use the same push model as we use for CI. We could have delivery pipelines get triggered on changes in our Git repository and then have the latest version of the manifests or helm charts applied to the target environment. In this chapter, we, however, focus on and favor the Pull model, which is implemented by GitOps operators or GitOps agents, such as Argo CD, Flux, and some commercial offerings.

GitOps operators in a nutshell – sync Git to K8s

GitOps operators continuously reconcile and ensure that the desired state declared in our repositories matches the actual state running in our target environments. The operator detects an *out-of-sync* if the state in Git does not match the one on the target environment. This could happen if the Git configuration was changed (e.g., a new image is available and causes a manifest update). It can also happen if somebody changes the configuration in K8s (e.g., through a manual update or a different automation tooling). The following is a high-level overview showing the role of the GitOps operator:

Figure 5.4: GitOps operator synchronizes the desired state in Git with the actual state in K8s

Depending on the GitOps operator tool, there will be different configuration options in the reconciliation process. It's worth checking out the documentation of the two prominent tools in the cloud native ecosystem: **Argo CD** *[9]* and **Flux** *[10]*. Also, make yourself familiar with the configuration elements, as some tools have a concept of projects and applications, whereas others just have a concept of a source.

Understanding reconciliation – keeping K8s in sync with Git

Essentially, the GitOps operator will fetch the desired state for a project or application from a source (Git repository, a folder in the Git repository, the OCI repository, S3 buckets, etc.) and compare it with the current state running on the target K8s cluster. The two states can either be *synced* or *out-of-sync*. When the state is *out-of-sync*, there are different options for the GitOps operator to synchronize the states – meaning: bring the current state to match the desired state:

- **Manual sync**: Either via a **command-line interface** (**CLI**) or a UI, one can trigger the GitOps operator to synchronize the two states.
- **Auto sync**: Once a system is out-of-sync, the GitOps operator tries to automatically synchronize.

- **Sync schedules/windows**: There might be times when you want or don't want syncs to happen. This is where sync schedules or sync windows come in, which either block or allow syncs to happen.

- **Sync failed**: It can happen that the GitOps operator can't apply the desired state. This could be because of configuration file mistakes (e.g., referencing an invalid image). It could be because of an infrastructure issue (e.g., K8s doesn't have enough resources). It could also be because of competing tools (e.g., auto-scaling tools, such as HPA/KEDA, changing replicas or resource limits). It's important to be aware of this state and handle it correctly. See the section on best practices for some additional input!

Now that we are aware of the synchronization basics, let's have a look at different GitOps operator deployment patterns!

GitOps operator patterns – single, hub and spoke, and many-to-many

The simplest version – and probably the model that many start with – is the monorepo approach with GitOps operators pulling into a single target environment, as we have seen in *Figure 5.3* for both patterns. While the single target is a common pattern, especially as you are getting started with GitOps, we have other patterns that we want to quickly highlight.

GitOps can also be set up where we have a central GitOps operator that keeps the desired state as declared in Git and synchronizes with several target environments by pushing out those changes. This is called the **hub-and-spoke model**. Another option is the **many-to-many model**, where each target environment has its own GitOps operator that continuously synchronizes the desired state with the state on its own cluster, as shown in the following figure:

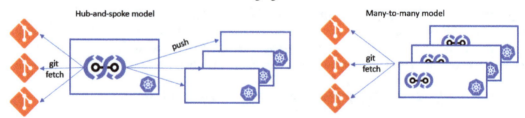

Figure 5.5: Hub-and-spoke and many-to-many GitOps operator pattern

Now that we have discussed the major phases in GitOps, let's recap and have a quick look into some best practices.

GitOps best practices

This list is not complete but should give you a good starting point when defining your GitOps process:

- **Separate config from source code repositories/folders**: It's recommended to separate the actual source code from the deployment definitions. Either in separate repositories or, within the repo, in separate folders. Why is that? It's a clean separation of concerns and access. It makes Git-triggered actions easier as a change in a deployment config file should trigger different actions than in the source code files. It avoids a potentially infinite loop of change if CI changes the same repository! The smaller the repository, the less work for GitOps tools to scan all files to determine the desired state.

- **Proper sync settings – poll versus webhook**: GitOps tools provide different sync settings for both scanning the source systems (e.g., Git) and scheduling the triggering of synchronizations. For scanning, make yourself familiar with the default poll frequency (e.g., Argo CD, by default, pulls all Git repos every three minutes). Both Argo CD and Flux can also be changed to receive webhooks from the Git system, which replaces the pull into a push mechanism! This is very important to understand, as with an increase in the number of source systems (Git, artifact repositories, S3 buckets, etc.), the number of API calls from your GitOps tool to those systems increases. It's a good practice to monitor the number of calls made from the GitOps tool to those external systems to get alerted in case the behavior drastically changes. A change in behavior could be caused by an accidental configuration change of default settings. Most tools provide Prometheus or OpenTelemetry metrics that can be observed by your observability tool!

  The authors have seen configurations that ended in API rate limits and even crashing Git systems due to too much load produced by the GitOps tools during a sync!

- **Not every config must be in a manifest**: As GitOps keeps track of desired versus actual state, it's important to leave some configuration out of your manifest that might be managed by different tools. Take replicas as an example. If you are using tools such as HPA or KEDA to auto-scale your pods, you do not want a static replicas count in your manifest. This would lead the GitOps tool to detect out-of-syncs for any change that HPA/KEDA does. This, therefore, results in two automation tools competing with each other.

- **GitOps notifications to handle sync states**: GitOps tools provide notifications when sync status changes. This would be when GitOps detects an out-of-sync, when it finishes a sync, or when there is an issue and a sync fails. In all those cases, it's important to get notified as you want to make sure that you handle sync failures or send information back to the development team when the latest update has successfully been synced.

To get notified, GitOps tools will create Kubernetes events that you can ingest into your observability solution and then react/alert on them. GitOps tools also typically provide some type of native notification feature where an external tool can be triggered in the case of a special event. Flux, for instance, provides Alerts, whereas Argo CD provides a concept of notifications. Both allow you to, for example, send Slack messages or trigger other external tools in the case of certain events that need attention!

> **GitOps – changing from push to pull**
>
> GitOps expands the power of Git from application code to everything as code. While CI/CD still focuses on building artifacts, GitOps provides an elegant way to pull the desired deployment state into any target environment along the software development life cycle. Changes to a system can only be done through Git with the benefit of traceability of changes, revertability to a previous version, and enforcing review processes through Pull Requests.

Now that we have covered the basics of GitOps, we should see how this can benefit us in building modern platforms. As platform teams, we can centrally enforce best practices (version control, policies, etc.) by using automation in CI/CD and the container registry to reduce the chance of a bad change request. With Git, every change and deployment is traceable back to a Git commit, making troubleshooting much easier, and it also provides an additional level of self-service (e.g., notify development teams when their change has been synchronized to the target environment or notify them when their latest version (via the `git commit` hash) has any issues).

Now, let's go on and spend some time on container registries as, without them, we wouldn't be able to publish or distribute any of the images that are being produced by the development teams that leverage the platform as a self-service.

Understanding the importance of container and artifact registries as entry points

Container and artifact registries deserve their own chapter as they are one of the core building blocks of modern cloud-native platforms. However, we will try to provide the relevant knowledge that should help you follow along with what's to come in the later chapters of this book.

We differentiate between public and private registries:

- **Public registries** are commonly used by individuals or small teams that want to get up and running with their registries as quickly as possible. However, at some point, it's worth looking at private registries.

- **Private registries** provide several critical capabilities, such as efficient storage of images, scanning for vulnerabilities, replicating images to other registries, enforcing access control when images get pulled, and notifying other tools about updates, with the ultimate goal of making images fast and easily accessible to those environments that need to deploy them.

While private container registries are typically only accessed internally to push new builds and have them pulled by our GitOps tools, we can also open the registry to the *public*. With the *public*, we mean allowing third-party vendors the option to push their latest images or deployment artifacts. As organizations rely on third-party software – think about any off-the-shelf software product you deploy yourself – we can leverage the same process of vulnerability scanning, replication, and access control before that software gets deployed on the internal systems. Financial One ACME, for instance, could allow their third-party vendors for a Development Ticketing System to push new versions to that public endpoint. Once scanned and validated, it can be deployed to the internal K8s clusters that run all the development tooling.

The following illustration is a very high-level overview of how container registries integrate into the end-to-end delivery process that starts with pushing a new artifact (third-party or CI/CD) until that new artifact gets deployed to the target environments:

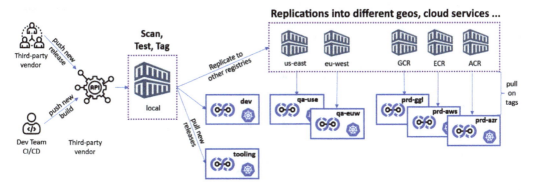

Figure 5.6: Container registries – the heartbeat of our platform

While there is a lot of in-depth information available from the different open source and commercial registry vendors, we want to give a quick overview of how registries fit into our platform engineering architecture, why certain concepts are important, and how we can best make container registries available to our end users as an easy to use self-service!

From container to artifact registry

Before diving into the process, we need to quickly highlight that container registries – while the name implies it – are not limited to container images. Most container registries typically support the OCI *[11]* image standard. Over the past years, container registries expanded to support non-container artifacts such as Helm Charts, zipped versions of Manifests, or Kustomize-based templates. That expansion also came with a name change for artifact registries as those tools manage artifacts in general and not just container images.

But that is not all. Artifact registries, open source, or **SaaS** (short for **Software as a Service**) services, often come with additional features, such as access control, regional replication, audit logging, policy enforcement, security scanning, notifications, and even more.

Building and pushing artifacts to the registry

Uploading (*pushing*) and *downloading* (*pulling*) images can be done using Docker (or other tools, such as Podman) commands. Before doing so, you need to authenticate against the registry (your private or public registry, such as Docker Hub). Once authenticated, it's easy to push and pull, as shown in the following code:

Interacting with the container registry via docker commands

```
export REGISTRY=registry.finone.acme

# Authenticate
docker login -u YOURUSER -p YOURPASSWORD $REGISTRY

# Build an image
docker build -t $REGISTRY/financeoneacme/fund-transfer:1.2.3 .

# Push an image
docker push $REGISTRY/financeoneacme/fund-transfer:1.2.3

# Pull an image
docker pull $REGISTRY/financeoneacme/fund-transfer:1.2.3
```

When building new images as part of the CI process, it's important to follow best practices around image labels and metadata. The OCI therefore also has a list of suggested annotations *[12]* as part of their image spec that should be used, such as `created`, `authors`, `url`, and `documentation` (all prefixed with `org.opencontainers.image.`).

Next to an API interface, registries typically also provide a **user interface** (**UI**) that makes it easier to see what images are uploaded, how much space they consume, and – depending on the management features of the registry – also provides the configurational aspects of all those capabilities (e.g., creating projects, managing users, specifying policies, configuring webhooks, etc.).

Managing uploaded artifacts

Registries typically manage artifacts in projects. Projects can be private or public within the registry, meaning that artifacts in a public project can be accessed by anyone who can reach the registry, while private projects can only be reached by authorized users. That brings us to user management and access controls, which can be defined on a project level, and can also feed into access logs. Registries will create logs for every push and pull to have a good audit trail. The CNCF Harbor is a very popular container registry that also provides good documentation about all those features. Instead of going into detail here, we suggest you read up on the publicly available documentation *[13]* for all those features. What is important to remember is how to organize your images into projects and who you give access to. If we also want to allow external third parties to push their images to our registries, you can create specific users for them that allow them to upload to the respective container projects!

Vulnerability scanning

Chapter 7 focuses on security in more detail but it's important to mention here that a central component, such as an artifact registry where every artifact has to pass through, is a perfect place for static vulnerability checking. Registries often provide out-of-the-box scanning capabilities or allow the integration of additional tools depending on the artifact type. Those tools would then be called either during the upload of a new image to get scanned as images arrive or on a schedule to make sure that images get scanned and rescanned on a continuous basis.

Subscribing to the life cycle events of an artifact in the registry

An artifact typically runs through different life cycle stages from the initial upload (push), security scan, replication to other repositories, and downloads (pull) until an artifact is deleted as it is no longer needed. Looking at Harbor as an example registry, we can also subscribe to all of those life cycle stages using webhooks. This enables a lot of interesting use cases, such as the following:

- **Security**: Notify security team on new vulnerabilities
- **Storage**: Clean up old images in case storage quotas are reached
- **Deploy**: New images uploaded from CI/CD or third-party vendors
- **Audit**: Keep track of who is pulling which artifacts

For reference, here is the full list of available life cycle events that can be used with webhooks: `artifact deleted`, `artifact pulled`, `artifact pushed`, `chart deleted`, `chart downloaded`, `chart uploaded`, `quota exceeded`, `quote near threshold`, `replication finished`, `scanning failed`, and `scanning finished`.

Retention and immutability

Repositories can grow really quickly – potentially leading to high storage costs if you don't have a good cleanup strategy. This is why it is advisable to clean old images that are no longer needed. Some registries, therefore, provide out-of-the-box retention policies where you can define the images that match certain rules that will be retained and for how long. Once images fall out of the retention period, they will be deleted.

Another use case is that, by default, everyone can upload a new image with the same image tag, leaving the previous image version tagless. To prevent this, some registries provide immutability rules, which prevent tags from being removed from existing images!

Monitoring our registries

As registries are the heartbeat of our platform, we need to make sure they stay healthy. If registries stop processing push or pull requests, can't execute vulnerability checks, and can't replicate to other registries or send out notifications, then we have a problem! It means that critical updates won't make it into the target environments fast enough. This could mean that a security vulnerability – while fixed through a new image – can't be remediated as the new image can't be delivered to the infected environment. It can also mean that we can't ship new features in the time we promised.

To observe the health of our registries, we can monitor their various health metrics. In self-hosted registries, those metrics are often exposed via Prometheus or a custom REST API. In SaaS-hosted registries, those metrics are typically exposed via the vendor's monitoring service API, such as AWS CloudWatch.

A good reference is Harbor, the CNCF project we mentioned earlier. It exposes lots of important metrics via **Prometheus** *[14]*, including the number of projects and repositories within a project, storage used, number of tasks executed and queued as well as performance metrics on the Harbor APIs (e.g., how long it takes to push or pull artifacts).

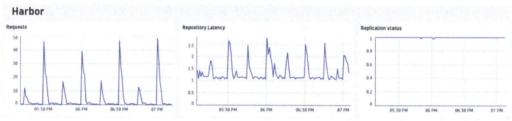

Figure 5.7: Monitoring our registry to identify potential problems early

Harbor is also an early adopter of OpenTelemetry, the CNCF project that initially introduced a standard for distributed tracing to the cloud-native community. Harbor provides an **OpenTelemetry exporter** *[15]*, generating traces with even more detailed information that can be used for both health monitoring as well as troubleshooting!

> **Registries – the central hub for all our artifacts**
>
> **Container or artifact registries** are the central hub of all software that gets built and deployed. It provides a central way to manage, scan, distribute, and enforce access to all software before it gets deployed. It must, therefore, be treated as a highly critical component that must be observed to ensure availability and resiliency, and must be secure in itself!

Now that we have talked about the importance of our artifact registry, let's continue to see how all of those components integrate with our end-to-release process.

Defining the release process and management

We have covered all of the building blocks it takes to build and deploy software. CI builds new container images or artifacts and pushes them to a registry. We know that registries have the power to scan for vulnerabilities, replicate to other registries, notify other tools about any activity, and enforce access control.

We also learned about GitOps and its pull approach of ensuring the desired deployment state (manifest files, Helm charts, Kustomize, etc.), as defined in a Git repository on the target environment.

What we are discussing now is how to define and enforce a full end-to-end release process and how to manage the life cycle of artifacts from the initial creation, initial deployment, promotion into other stages, and updates to new versions until a potential retirement when the software is no longer required!

The following illustration highlights that pushing a new image and replicating it to other registries is just one piece of the puzzle. The new version of the image also needs to be referenced in the deployment definition that is managed in mono or poly Git repositories. When having multiple stages (development, QA, production) and having multiple regions or clusters in a stage, we also need to promote that new version between each stage and each region/cluster!

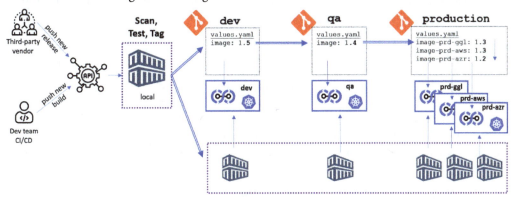

Figure 5.8: Release process – from initial push to deploy and promote, from stage to stage

In the preceding figure, `values.yaml` represents the values for a Helm chart. It is, of course, simplified as there would be many more of the values that we discussed earlier (e.g., ownership, application context, log level, `git commit` hash, container registry, etc.).

Now, let's have a look into how those deployment updates can be done, what should happen between the stages, what rollout options we have, and how to best keep track of a live inventory to know what is deployed where, when, and by whom!

Updating deployment to a new version

There are different ways that `values.yaml` can be updated for the initial development stage and how it then gets promoted to the next stages. Depending on the tooling and the level of maturity or processes that organizations have in place, this can be implemented in various ways. The options would be as follows:

- **Updated as part of the CI**: Once the CI publishes a new image to the registry, it could open a Pull Request to update the version in the repository for the first stage.

- **Updated through a registry webhook**: When a new image is uploaded to the registry, a webhook can be used to open a Pull Request. This approach is similar to having the CI do it. However, it decouples the process. This also works independently of which tool (e.g., CI) or creator (e.g., third-party vendor) pushes a new image version.

- **Scheduled updates**: Every hour, every day at 8 a.m., or any other schedule can be used to create a Pull Request with the latest version information from the registry. This provides continuous updates as well as batch changes on a schedule.

- **Manual Pull Requests**: As a precursor to automating the Pull Requests, or to enforce mandatory manual approval, the version update can also be done manually.

Batching changes to combat dependencies

When dealing with simple applications or microservices that can be deployed independently, a single file update might be enough. Often, we have to combine multiple changes into a single change request. This is when changes have dependencies on changes in other components. An example for our Finance One ACME could be that the new version of the `fund-transfer` service also requires a new version of the `account-info` service, which requires an infrastructure change. You can see that it can easily become complex and it becomes harder to fully automate.

Those changes are hard to automate and often fall back to some release management team that resolves those dependencies manually.

There are other ways of solving this. One approach is using package managers, such as Helm, where a Helm chart can contain all configurations needed to deploy a full app. Helm charts can then also become artifacts uploaded to a registry.

Another approach that we have seen in an earlier section is using tools such as Crossplane, which provides IaC and application as code. The following is an example of using a composition for an application that contains several components, such as the business logic, a cache, an ingress, and a database:

```
apiVersion: composites.financialone.acme/v1alpha1
kind: FinancialBackend
metadata:
```

```
    name: tenantABC-eu-west
spec:
  service-versions:
    fund-transfer: 2.34.3
    account-info: 1.17.0
  redis-cache:
    version: 7.4.2
    name: transfer-cache
    size: medium
  database:
    size: large
    name: accounts-db
  region: "eu-west"
  ingress:
    url: "https://tenantABC-eu-west.financialone.acme"
```

As we can see, there are different ways to solve this application dependency problem.

Where this becomes trickier is when we have cross-application dependencies or dependencies to shared services that are independently deployed and updated. Following best practices to define good and clear APIs between those services and ensuring backward compatibility between major versions will make this easier. To learn more, have a look at existing literature or the work the CNCF **TAG App Delivery** *[16]* does on managing application and deployment dependencies.

Pre- and post-deployment checks

Once we have all our deployment configuration changes ready and committed to Git, we will be ready to deploy into the target cluster. As explained earlier in this chapter, we can either use a Push model (e.g., a delivery pipeline deploys our changes) or a Pull model (e.g., a GitOps operator synchronizes the latest version in Git to the target cluster).

With both approaches, we want to do some pre- and post-deployment checks to answer questions, such as the following:

- **Pre-deployment: Are we ready to deploy that change?**

  - Are all external dependencies ready?

  - Have all new images been successfully scanned and have no vulnerabilities?

  - Is there no ongoing maintenance or deployment quarantine in the target environment?

  - Has everyone that needs to approve a deployment approved it?

- **Post-deployment: Was the deployment successful?**

  - Are updated services available and successfully handling requests?

  - Does the system meet all functional and non-functional requirements?

  - Do all services meet their SLAs and **service-level objectives (SLOs)**?

  - Was everyone notified about the deployment?

  - Can the deployment be promoted to the next stage?

The pre-deployment checks can be implemented in various ways. Some can be implemented as part of the Pull Request flow (e.g., Pull Requests cannot be merged when not all pre-deployment checks are fulfilled). They can also be validated within a deployment pipeline or implemented as a pre-sync hook in GitOps tools.

The post-deployment checks can be implemented after the deployment is done from the deployment pipeline as well as through post-sync hooks in GitOps tools.

One tool that enables pre- and post-deployment checks on Kubernetes independent of how the deployment is done is the CNCF project, Keptn. **Keptn** can stop Kubernetes deployments if the pre-deployment checks are not successful. Keptn can also execute post-deployment checks, such as executing tests, evaluating SLOs, or notifying teams.

To learn more about Keptn, check the documentation *[17]*.

Deployment notifications

Once we are aware of a new deployment that has either succeeded or failed the post-deployment check, we can use this information and notify those people or tools that can benefit from this information. Whether using tools such as Keptn, the notifications of the GitOps tools, or doing this from the pipeline, here are some examples of where to send this information:

- **A chat**: Notify development teams in their Slack channel that their latest deployment is either ready or has failed the checks

- **A ticket**: Update the Pull Request or a Jira ticket with the information about the deployment status

- **A status page**: Keep a deployment status page with version, environment, and health information up to date

- **An observability backend**: Send a deployment event to your observability backend

> **Deployment events to observability backend**
>
> Observability tools not only track metrics, logs, and traces, but they can also track events such as deployment or configuration change events. Many observability tools (**Dynatrace**, **Datadog**, **New Relic**, etc.) can use those events and correlate them with a change in the app's behavior. This can significantly improve incident response as it can be correlated with a specific deployment change.

Promotions between stages

Assuming we have a new release in development that successfully passed all post-deployment checks, how do those changes get promoted from development to QA and then into production? Does every change have to get promoted all the way into production or not? Do new releases get rolled out to all production environments at once or are there better strategies for it?

Here are some strategies that we have seen in organizations we have worked with in the past. For all those strategies, it means that a Pull Request is created to promote the new deployment definition from the Git location of the lower stage to the Git location of the higher stage:

- **Automated**: Often seen between development and QA environments. Every time the post-deployment checks are successful, it can trigger an automated Pull Request. This ensures that all development changes that have made it through the basic checks are quickly promoted to an environment used for more thorough testing.

- **Scheduled**: For instance, once a day, promote the latest version from development or QA into a special test environment (e.g., a performance testing environment). This ensures that the team gets daily feedback on performance behavior changes of their updates from the previous 24 hours.

- **Controlled/manual**: This typically happens the closer you get into production. Changes that made it successfully through QA and performance testing are marked to be safe to promote into higher-level environments. The actual promotion typically happens manually while the Pull Request itself might already be automatically created (but not auto-approved) from those versions that successfully made it through, for example, the performance testing phase!

- **Multi-stage in production**: When having multiple production clusters, it is common practice to roll out the changes into one cluster first. Then, validate if everything works and keep rolling out the rest. This first cluster could be for internal usage only or for users who know that they receive updates first (e.g., friends and family or members of an early access group). When deploying into multiple regions or multiple SaaS vendors, it is also advisable to define a clear sequence (e.g., start in Europe first and then roll out to the US).

Using multiple quality gated stages and a staged rollout strategy in our various production environments has one goal: reducing the risk of failed deployments!

Blue/green, canary, and feature flagging

While using multiple stages with pre- and post-deployment checks already reduces risk, there is more we can do for each individual deployment: **progressive delivery** strategies, such as blue/green, canary, or feature flags. We have already discussed what those things are in more detail in the *Continuous deployment – decoupling deployments from releases* section.

In the context of the release process, this is an important topic, as there are several open questions:

- Who is receiving the new version? Is it a certain percentage of users or a specific group?
- How do you measure and validate whether the rollout was successful?
- Who is responsible for making the roll-forward or rollback decision?

Just like validating a deployment through post-deployment checks, we can do the same with progressive delivery. Open source tools, such as **Argo Rollouts** *[18]* and **Flagger** *[19]*, or commercial tools provide automated analysis between the progressive rollout phases. This usually works by querying data from the observability platform; for example, Prometheus, to validate if the new version doesn't differ from the existing version for key service-level indicators, such as request failure rate, response time, memory, and CPU consumption. For more details, it's recommended to check the documentation of the respective tooling that is used for progressive rollouts.

Release inventory

The value proposition of automating the deployment and release process is that engineering teams can release updates more frequently into the various target environments. The easier we make this automation available through our platform, the more teams will end up deploying more releases.

Release inventories allow us to keep track of which versions are currently released in which environments by which teams. From a platform engineering perspective, we can enforce a consistent definition of exactly that information: version, environment, and ownership. If everything is configured as code, it means that this information is available in Git. When using Kubernetes as the target platform, we can also add this information as annotations on our Kubernetes objects (Deployments, Pods, Ingress).

In previous examples, we already highlighted some of the standard K8s annotations. Here is a snippet of a deployment definition:

```
apiVersion: apps/v1
kind: Deployment
metadata:
  name: fund-transfer-service
spec:
  ...
  template:
    metadata:
```

```
annotations:
   owner.team: team-us
   app.kubernetes.io/name: fund-transfer
   app.kubernetes.io/part-of: backend-tenant-2
   app.kubernetes.io/version: 1.5.1
```

With this information in K8s, we can simply query the K8s API of each K8s cluster to get an overview of all deployments by those labels. We could then get an overview like this:

Cluster	Release	Part-Of	Owner
dev	fund-transfer:1.5.1	backend	team-at
dev	account-info:1.17.2	backend	team-ae
qa	fund-transfer:1.4.9	backend	team-at
qa	account-info:1.17.1	backend	team-ae
prod-eu	fund-transfer:1.4.7	backend-tenant-1	team-de
prod-eu	account-info:1.17.1	backend-tenant-1	team-de
prod-us	fund-transfer:1.4.6	backend-tenant-2	team-us
prod-us	account-info:1.17.0	backend-tenant-2	team-us

Table 5.2: Release inventory based on metadata on deployed artifacts

Observability tools often provide this feature as they already pull the K8s APIs to observe cluster health, events, and objects including this metadata.

Another approach is to parse this information directly from Git. Tools such as Backstage do exactly this and with that information, provide an easy-to-search software catalog.

Release management – from launch to mission control

For many organizations, the release process ends with the actual deployment of a new release into production when operations teams take over all aspects of keeping the software running. These two phases are also often called **launch** and **mission control**, borrowed from how NASA manages their space missions.

Many years ago, the DevOps movement was fueled by prominent examples, such as the AWS platform, which promoted the *you built it, you run it!* approach. This meant that the responsibility of developers didn't end with building an artifact and then throwing it over the so-called "Ops Wall." Developers had to take full responsibility and ownership of their code from development all the way into production. They had to handle incidents and updates until their code eventually got retired.

When looking at today's complex environments, it is very hard to own every aspect (code and infrastructure). With platform engineering, we try to bring the promise of DevOps back by providing self-services to reduce the complexity and giving teams the chance to own more of the end-to-end life cycle of their software.

When building future platforms, our focus must be to provide self-services to orchestrate the whole life cycle of an artifact. In fact, orchestrating the whole life cycle of an application as a single artifact is typically just a fraction of an application. **Life cycle orchestration** includes building, deploying, and releasing, but also all use cases needed for production support. This includes resiliency through scaling, incident management, access to the right observability data for troubleshooting, and automated delivery to push fixes and updates.

In the next section, we dive into life cycle orchestration, how to increase transparency by making our life cycles observable, and how an event-driven model can help reduce the complexity of pipeline and orchestration code.

Achieving sustainable CI/CD for DevOps – application life cycle orchestration

From building, to publishing, to promoting, to deploying, to releasing, to fixing. That's a lot of tasks that have to be automated to orchestrate the whole life cycle of our artifacts and applications. The challenge that we have seen with teams that take end-to-end responsibility is that the majority of those scripts deal with the following:

- Triggering a certain (hardcoded) tool

- Doing so in a certain (hardcoded) environment

- Then, waiting for the tool to complete its job

- Parsing the specific result format

- Based on the parsing result, deciding what to do next (hardcoded)

That script code then often gets copied and pasted between different projects, slightly modified, and adapted. Fixes or adaptations to a single script are hard to promote to all the other variations as there is no easy tracking of all those scripts. This leads to high effort in maintenance, makes it inflexible to change the process or tools used in those scripts, and takes away time from engineers doing their regular jobs.

There is nothing wrong with using the power of pipeline scripting. Many of the pipeline automation tools also provide abstractions, code reuse through libraries, and other features to reduce code duplication and increase reusability. So, depending on the tools you use, make sure to follow those best practices!

There exists a different approach that fits in the event-driven nature of modern cloud-native applications and cloud-native environments: **life cycle event-driven orchestration**!

In the next section, we will dive into this approach as it provides a lot of flexibility and centralized observability and will lead to a more sustainable way to automate CI/CD and operations!

Artifact life cycle event observability

The basics for event-driven orchestration are life cycle events, such as those discussed in *Chapter 3*. It's about observing the full life cycle of an artifact: from the initial Git commit of code, the building and pushing of container images to a registry, to the releasing into every stage until the artifact gets updated or retired, including all the steps in between.

So far in this chapter, we discussed many of those life cycle steps and also highlighted how to extract some of those life cycle events:

- **CI/CD pipelines** can emit events when they *build* or *deploy*

- **Artifact registries** provide webhooks when containers get *pushed*, *pulled*, or *scanned*

- **GitOps** tools can send notifications for `PreSync`, `Sync`, `PostSync`, and `SyncFailed`

- **Git** workflows can be used to send events when code *changes* or gets *promoted*

- Tools such as **Keptn** provide pre- and post-deployment events for individual deployments as well as complex applications

- **Container** platforms, such as Kubernetes, expose events about *deployment health*

CDEvents *[20]*, a project from the Continuous Delivery Foundation, extends the CloudEvents specification, which is already a graduated CNCF project with wide ecosystem adoption. CDEvents was initially started to standardize events for all life cycle phases for building, testing, and deploying. It has recently expanded to also cover the life cycle phases of deployments in operations such as production incidents.

The idea is that tools that adhere to those event standards are easier to integrate with all other tools in the ecosystem. Instead of having to manage and maintain hard code integrations between tools, those tools communicate via open standard APIs and Events. Tools can emit those events; for example, Jenkins has created a new artifact and other tools can subscribe to them (e.g., GitLab can subscribe and trigger a workflow to scan and publish the container). That life cycle phase would also generate a standardized event.

All events have a minimum set of properties to identify the phase, the artifact, and the tool that was involved, as well as the additional properties that are mandatory (e.g., initial `git commit`, version, environment, and responsible team).

When all those events are sent to a central event hub, it enables a fully event-driven orchestration of the life cycle of artifacts.

For example, if you decide to notify development teams about new deployments in their Slack, you can easily subscribe to the deployment events and forward that event to Slack. If you change the chat tool to something else, simply change that event subscription. No need to find all code in all pipelines that currently send notifications as this happens through a simple event subscription change.

To sum it up, using a well-defined set of life cycle events across all tools and phases enables many capabilities in our platform engineering approach:

- **Traceability**: We can trace every artifact from its initial creation until its end of the life cycle. This allows us to see where artifacts are (release inventory), where they are stuck, and who is responsible for letting an artifact into a certain environment.

- **Measurement**: We can measure how many artifacts flow through the life cycle and how long it takes. This is the basis for reporting the DORA Efficiency metrics!

- **Interoperability**: We can easily integrate new tools or replace them if they all adhere to the same standards (e.g., switching from one notification tool to another is just a matter of changing an event subscription).

- **Flexibility**: Like replacing tools, we can easily adapt our delivery processes by having additional tools add work on certain events (e.g., adding an additional mandatory security scan for deployments in certain environments can be done by an event subscription of a security scan tool).

Let's have a quick overview of the building blocks of such an event-driven system as compared to having to create and maintain lengthy complex automation scripts to include a lot of process logic.

Working with events

The first step is that the tools we use along the artifact life cycle emit those events. Looking at CDEvents (which extends CloudEvents), there are several tools in the ecosystem that already provide out-of-the-box support for them, such as Jenkins, Tekton, Keptn, Tracetest, Spinnaker, and others.

Those tools that are not integrated yet can easily emit those events using the available SDKs. The following is a code example using Python, which creates a Pipeline Run Finished Event (`cdevents.new_pipelinerun_finished_event`) with metadata, identifying the pipeline, artifact, or owner:

```
import cdevents
event = cdevents.new_pipelinerun_finished_event(
  context_id="git-abcdef1231",
  context_source="jenkins",
  context_timestamp=datetime.datetime.now(),
  subject_id="pipeline_job123",
  custom_data={
    "owner": "dev-team-backend",
    "part-of": "backend-services",
    "name" : "fund-transfer-service"
```

```
  },
  subject_source="build",
  custom_data_content_type="application/json",
  pipeline_name="backendBuildPipeline",
  url="https://finone.acme/ci/job123",
)
# Create a CloudEvent from the CDEvent
cloudevent = cdevents.to_cloudevent(event)
# Creates the HTTP request representation of the CloudEvent in
structured content mode
headers, body = to_structured(event)
# POST it to the event bus, data store
requests.post("<some-url>", data=body, headers=headers)
```

The preceding Python code example creates a new `pipelinerun_finished_event`, which indicates the finished execution of a pipeline. The additional context data indicates which pipeline and when it was built, and it allows us to provide additional metadata, such as ownership, artifact, or which application this pipeline belongs to.

Whether you use the CDEvents standard proposal or send your own life cycle events, it is a good idea to base it on CloudEvents as that project already has a lot of industry integrations that can either emit or consume CloudEvents, with Knative as one example!

Subscribing to events to orchestrate

Once all our tools are emitting standardized events, we can more easily orchestrate our artifact life cycle process by having tools we want to participate in the process subscribe to those events they want to act upon.

We discussed the same concept earlier in this chapter when we talked about using the webhook capabilities of tools such as Argo CD or Harbor to act upon when a new artifact is available or when a new deployment was successfully synced.

The benefit of a standard event model is that tools no longer need to subscribe to a specific webhook of a specific tool (e.g., Argo CD webhooks). Instead, we can subscribe to a central event hub that receives all standardized life cycle events from all involved tools, as visualized here:

Figure 5.9: All tools emit and subscribe to standardized life cycle events

Having everything based on an event standard eliminates the need for point-to-point tool integrations or hardcoded tool integrations in pipeline scripts. As an example, instead of sending a notification to Slack or Mattermost from every pipeline that deploys a new build, we can simply subscribe to the `dev.cdevents.service.published` event and have the details about that service forwarded to our chat tool.

If that tool is Slack today, and at a later time, we decide to move to another tool, we simply change that subscription.

Another use case is to have different tools as part of the process active in different environments. As those standardized events contain a lot of metadata (e.g., which environment a service gets deployed into), we can subscribe to events for a certain environment. The following is a table that shows some examples:

Source Tool	Event Properties	Subscribed By	Action
Argo	Type: `service.deployed` Environment: `staging` Artifact: `fund-transfer:2.3` Owner: `team-backend`	K6 when `environment == "staging"`	Execute simple load
.		Slack when `Owner == team-backend`	Send a notification to the team's backend Slack channel
		OTel Collector	Collecting all events to forward to the observability backend

Table 5.3: The same event from Argo can be subscribed by various tools for various actions

Some tools already provide out-of-the-box support for CloudEvents where they can either subscribe to a CloudEvent source or provide an API endpoint that can consume CloudEvents. For others, it will be necessary to build a slim integration layer where one subscribes to those events and then forwards them to the target tool. It's also possible to implement this using event bus systems that support CloudEvents.

Analyzing events

Now that we know how events are sent and how they can be subscribed by other tools, we can discuss how we can leverage them to analyze how well our life cycle processes actually work.

Well-defined events that have a timestamp, a life cycle phase definition (=`event type`), and some context (artifact, environment, owner) can be analyzed to answer questions, such as the following:

- How many deployments happen in a given environment?
- How many deployments are done for a particular application or tenant?
- How active is a team based on the ownership information?
- How many artifacts make it from development to production?
- Which artifacts take a long time and where are they blocked on the way to production?
- Are there certain artifacts that cause more security vulnerabilities or production problems than others?
- Which tools involved in the process are consuming most of the time?
- Are there tools that are most often the reason for a slow end-to-end process?

There are probably many more questions that we can all answer by analyzing those events.

How can we analyze them? You can stream all those events to a database or your observability platform. In *Figure 5.9*, we included OpenTelemetry, as events can just be ingested and forwarded to your observability backend and analyzed there.

> **Bringing transparency into CI/CD through event observability**
>
> Having all events in a single spot with all that metadata and clearly defined types that represent the life cycle stages allows us to get a lot of transparency into the integration, delivery, and operations processes. This data allows us to optimize our processes, which will result in more sustainability.

Building automation for CI/CD and operations typically results in a lot of customized code that needs to be maintained across all projects it was copied to. What we learned in this section is that moving to an event-driven approach for artifact life cycle management can address a lot of the complexity problems that are otherwise hidden in custom scripts or hardcoded tool-to-tool integrations.

In *Chapter 9*, we will cover additional aspects of how to reduce technical debt in all our platform components by making the right architectural decisions.

IDPs – the automation Kraken in the platform

In this chapter so far, we have learned a lot about the basic building blocks to automate the end-to-end build, delivery, deployment, and release process. We have talked about new approaches to deploying the desired state with GitOps where the desired state is pulled from within the target environment versus pushed from an external tool, such as a pipeline.

We discussed the end-to-end release processes on what happens from the first commit until releasing software to the end users. Finally, we talked about applying an event-driven approach to orchestrating our artifact life cycle, which provides a centralized event hub to make everything that happens more transparent and observable. It also gives us more flexibility as we can remove the complexity of tool integrations and process definitions from pipeline or bash scripts into event subscriptions and event-driven workflows.

In this last section, we want to have a brief look into which of those concepts can be implemented with existing tools that you may already have, which new approaches exist to solve some of the challenges we discussed, and where you may want to look as some new tools have emerged over the last years – both open source and commercial – that take some of that work off our shoulders.

We will do so by putting ourselves into the shoes of our users, our developers, or our development teams, as they are the ones who will need to apply to a new way of working.

Providing templates as Golden Paths for easier starts!

We try to enforce a lot of new practices, such as ensuring the right metadata on the deployments (version, application context, ownership, etc.) or having security vulnerability checks as part of every build pipeline. In the platform engineering community, those are also referred to as *Golden Paths*. To ensure that those practices can easily be followed by teams that start new projects, we need to make them easily accessible and adoptable!

The easiest and most impactful approach is to provide software or repository templates. These are templates in the forms of manifest files, pipelines, automation scripts, and so on that developers can find in a template repository, which they can then take and apply to their projects.

While this approach works, it doesn't force engineers to really use those templates; plus, it's an additional manual step that can also lead to mistakes.

One way to make this easier and automated is to either provide a CLI or a UI to initialize new or update existing git repositories with best-practice templates. This can either be custom-built or we can look into existing solutions, such as **Backstage**, a CNCF project that was donated by Spotify.

Backstage's **Software Templates** feature was built to make Golden Path templates the entry point for every developer's journey as they are building new software components. Templates can be defined by subject matter experts who know how to properly configure pipelines, enable automated testing and deployment, and enforce security checks.

Once templates are defined, they are available through an easy-to-use wizard that prompts the developer for some critical input data, such as what type of service they implement, ownership information, the requirements on observability or security, and so on – all of that input will then impact the creation of a new repository or the update of an existing one with the files and configurations from the template.

To learn more about templating, check out the detailed documentation and examples on the **Backstage website** *[21]*.

Abstractions through Crossplane

Another simplification and way to enforce best practices is to provide an additional layer of abstraction when defining your application or services. In K8s, we have to define our deployments, services, Ingress, **Persistent Volume Claims** (**PVCs**), and even more when we need to deploy dependent services, such as a database, a cache, or any other required software components.

In the earlier section on IaC, we introduced the CNCF project, Crossplane. **Crossplane** orchestrates both infrastructure and application deployment through code and provides a concept of so-called **composites**. We will not spend more time here on this as we already provided several examples earlier on how to use composites to provision a performance test environment as well as one to define a financial backend type of application where the developer only needs to specify the versions of services that will then be deployed together. The following are just the first two lines of that composite definition. See the rest in the *Crossplane – IaC for platform and applications* section:

```
apiVersion: composites.financialone.acme/v1alpha1
kind: FinancialBackend
```

When providing abstractions, it is important to make them known to developers. This can be done by providing educational material or simply providing them through the same templating approach, as discussed earlier, using a tool such as Backstage.

Everything Git-flow-driven

Well, it should be no surprise that Git is our source of truth – we have established this early in the chapter. However, most Git solutions provide additional capabilities that we can use to also enforce standards and processes (e.g., GitHub workflows). Workflows can be triggered on a schedule or as part of many different events that can happen in the end to end flow of a Git driven process (e.g., *push, pull requests, release*).

This allows us to enforce our standards as well before artifacts get built and pushed; for instance, validating mandatory metadata files we expect for every deployment (e.g., ownership information). We can also use this to automatically do code scans and generate scorecards or we can use it to validate that all dependencies are safe and don't have any known security vulnerabilities.

Depending on the Git tool that is chosen, you will typically find a marketplace or best practice catalog of workflows and actions that can and should be executed for certain types of projects. Make sure you make yourself familiar with all that is possible based on your tool choice.

Software catalog

Once we enable developers to build more software that follows all our processes, we will hopefully see the result in a lot of new services being developed. Those are services that other developers also need to know about so that we avoid the problem of development teams building duplicated services and encouraging developers to build more capabilities on top of existing services and APIs.

A **software catalog** that gives an overview of all available services and APIs and ideally also provides some documentation is what we are aiming for.

As Git is the source of truth, we can extract most of this information straight from Git. Depending on which Git solution we choose, a software catalog might already be part of the offering. However, there are more services and APIs that are part of the software catalog that an organization owns and can develop against (e.g., external APIs or third-party software deployed on-premises).

Backstage, the tool that also provides the templating feature discussed earlier, also comes with a software catalog. It gets its data from parsing specific metadata files in Git repositories but also allows external data sources to provide entity information. The following illustration is taken from the Backstage blog and shows what Spotify's software catalog in Backstage looks like:

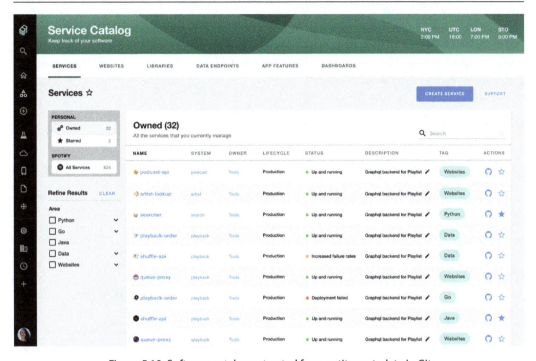

Figure 5.10: Software catalog extracted from entity metadata in Git

As we can see from the preceding screenshot, software catalogs are a powerful way to understand what software components are available within an organization, what type of software it is, who owns it, where to find the source code, and additional information.

Tools such as Backstage are not the full IDP; however, they represent a portal – a graphical UI – into all the data relevant for the majority of the users of an IDP.

While Backstage is one option, there are many other options out there. Everything from homegrown to other open source or commercial tools, such as Cortex, Humanitec, Port, or Kratix.

Summary

In this chapter, we learned a lot about the underlying automation and processes to get an artifact from the initial creation all the way into production. For modern platforms, a GitOps approach where we pull versus push changes should be a key consideration. We learned about Git as the source of truth and artifacts (container or OCI-compliant images) as of our business logic into our target environments.

As an organization grows, it's important to enforce good processes and best practices. For enforcement to work, it needs to be easily accessible and should be available end-to-end as a self-service to not impact the flow of creativity of engineers.

This also brings us to the topic of the next chapter. In *Chapter 6*, we dive into the importance of focusing on self-service capabilities that really address the needs of our target users: our developers. We will discuss how to bring those Golden Paths' best practices into our platform to significantly improve the way developers can get their work done.

Further reading

- [1] OpenFeature – https://openfeature.dev/
- [2] Crossplane – https://www.crossplane.io/
- [3] Renovate Bot – https://github.com/renovatebot/renovate
- [4] Semantic Versioning – https://semver.org/
- [5] Kustomize – https://kustomize.io/
- [6] Helm – https://helm.sh/
- [7] *The Pragmatic Programmer* – the DRY principle – https://media.pragprog.com/titles/tpp20/dry.pdf
- [8] *How to set up GitOps directory structure* – https://developers.redhat.com/articles/2022/09/07/how-set-your-gitops-directory-structure#directory_structures
- [9] Argo CD – https://argo-cd.readthedocs.io/en/stable/
- [10] Flux – https://fluxcd.io/flux/
- [11] Open Container Initiative – https://opencontainers.org/
- [12] Suggested container annotations – https://github.com/opencontainers/image-spec/blob/main/annotations.md
- [13] Harbor – https://goharbor.io/
- [14] Harbor Prometheus metrics – https://goharbor.io/docs/2.10.0/administration/metrics/
- [15] Harbor distributed tracing – https://goharbor.io/docs/2.10.0/administration/distributed-tracing/
- [16] *CNCF TAG App Delivery* – https://tag-app-delivery.cncf.io/
- [17] Keptn – https://keptn.sh/
- [18] *Argo Rollouts* – https://argoproj.github.io/rollouts/
- [19] Flagger – https://flagger.app/
- [20] CDEvents – https://github.com/cdevents
- [21] Backstage – https://backstage.io/

6

Build for Developers and Their Self-Service

A platform without accessible features is not a platform. As a rule, self-service should be included and must improve accessibility to add value. In the process of building platforms, there is often no distinction between infrastructure and application layers. The integration of **Internal Developer Portals (IDPs)** adds new complexity to this, addressing some demands while leaving others unanswered. It is important to bring the community into your platform and be open to their contributions.

In this chapter, we will review some concepts and share some ideas for best practices. By the end of the chapter, you should understand how to approach the building of a platform that is resilient, flexible, and meets your users where they are.

In this chapter, we'll cover the following main topics:

- Software versus platform development – avoiding a mix
- Reducing cognitive load
- Self-service developer portals
- Land, expand, and integrate your IDP
- Architectural considerations for observability in a platform
- Open your platform for community and collaboration

Technical requirements

In this chapter, there will be some technical examples with `.yaml` files and commands. While you don't need to set up a cluster to follow along, doing so may enhance your understanding. We used the following technologies to develop our samples and explanations:

- kind – the version tested was kind `v0.22.0 go1.20.13`
- We used this guide to set up a three-worker node cluster: `https://kind.sigs.k8s.io/docs/user/quick-start/#configuring-your-kind-cluster`
- Docker (a Docker Rootless setup is recommended)
- The `kubectl` command-line tool
- A GitHub repo

The code examples can be found inside the `Chapter06` folder here: `https://github.com/PacktPublishing/Platform-Engineering-for-Architects`.

We'll do a couple of small tutorials over the course of the chapter. While not every code snippet needs to be run against a Kubernetes cluster, it is recommended to set up a local Kind cluster with at least one control plane node and three worker nodes to get the full value of the tutorials. The configuration for the Kind cluster can be found in the GitHub repo for the chapter.

Software versus platform development – avoiding a mix

In our journey toward platform development, we have to disambiguate between this and other forms of software development. Remember that the purpose of the platform is to enable development and operations teams; it is not something that customers directly experience, although they do benefit indirectly.

The value of the platform is the unification of all the tools, services, and applications required to build and land an application in front of users. In short, a platform is a series of services used by developers to deliver software applications into the hands of end users. You can consume existing software applications, so it's possible to develop a platform without writing any code.

So, how do you develop a platform, and how do you develop software? Where do they intersect, and where do they differ?

The platform life cycle versus the software life cycle

In many regards, the life cycle of a platform looks similar to any **Software Development Life Cycle (SDLC)**.

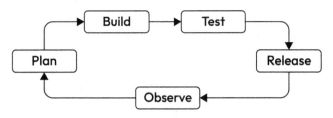

Figure 6.1: An SDLC

In the previous chapters, we covered the benefits of a robust planning phase and treating your platform as a product. And while it's important not to over-engineer the platform during this process, you'll still end up with a system that has a variety of aspects. Thus, due to its relative size and complexity, the release of a platform may not be as simple as pressing a button and letting your change propagate through some servers or clusters. This is in no small part due to the number of moving pieces you can find within a platform. For example, creating a new integration to deliver a **Software Bill of Materials (SBOM)** will be less disruptive to users than adjusting policies within an existing policy engine to be more restrictive.

If we look at the anatomy of a typical platform and its user considerations, it will be easier to understand how the platform life cycle differs from a software application.

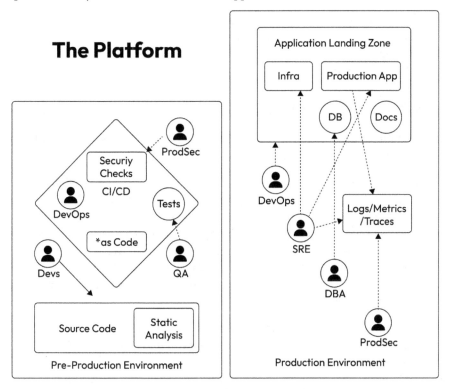

Figure 6.2: An example of IDP components and functional areas

In this platform example, pre-production is on the left, and the production environment where customer-facing applications will be deployed is on the right (**Application Landing Zone**). Whether a single Kubernetes cluster, multi-cluster, or multi-architecture, the platform is a cohesive unit that encapsulates these two paradigms. The platform should abstract the architecture away from users, allowing them to work independently.

Here, DevOps engineers care about how the application is built and released. Things such as progressive rollouts, DORA metrics (`https://dora.dev/`), and other DevOps practices are what the platform may be expected to support. Similarly, if your organization is quite far along in its cloud-native journey, the data surrounding the operations of the production application would be something site reliability engineers would concern themselves with. And there are other relevant considerations, such as security checks, infrastructure as code, and docs. Each aspect of these can be controlled individually within a platform and made available to all or a subset of the personas.

Figure 6.2 also assumes there are personas working in dedicated roles; however, if there's no dedicated DevOps team, it's likely that developers and quality engineering would share responsibilities in that space. The overall concept should still apply even to small organizations, where one person may have a stake in multiple personas' user stories. Even the theoretical division of the personas should assist when undertaking the life cycle of a platform.

A platform is more than just the software; it's also its tuning in concert with the workloads that help make sure it's performant for its users. While the needs of the users do influence the size, scope, and functionality of the platform, it should be designed as agnostically as possible so that one user's golden path isn't prioritized over another.

Reliability versus serviceability

A platform's **reliability** and **serviceability** are the largest factors of its usability by developers in an organization. But what do each of these things really mean?

At its core, reliability encompasses the overall availability of a platform and the ability of a user to successfully interact with it. Reliability in a platform should assume that multi-tenancy (multiple users who are fully isolated from each other) has been implemented in the platform. The reason for this is that the concept can still be applied where it's not required, but your platform still supports multiple users with individual needs. While multi-tenancy isn't an intuitive approach for an internal tool, highly regulated teams may prefer to work in an isolated environment to ensure that there's no risk of security and compliance violations. Even if all your users are internal customers, they still shouldn't be exposed to or impacted by other user workloads.

To attain the reliability of a platform, you should leverage tuning and a policy that helps you ensure workload isolation and maintain the integrity of a platform. In a Kubernetes environment, a node may be a virtual machine or real hardware. As with any machine, each node has a set amount of memory and CPU allocated to it. However, unlike your standard virtual machine, when you overcommit the CPU, a node can use the CPU that would otherwise belong to a different node. This is not true of memory, although both can be oversubscribed, and **Out of Memory (OOM)** will cause a node to reboot. This allows the overall resource requirements to be kept lower by guaranteeing a resource will be able to utilize CPU cycles if they're available; however, on the other hand, a bad process that eats CPU due to a software error may never return the CPU cycles to the node pool for availability. Other factors, such as Pod priority, requests, limits, and Pod disruption budgets, can influence how a scheduler will determine where and how to schedule Pods, and whether or not the kubelet will evict an existing workload. For more information on these topics, it's important to review the Kubernetes documentation regularly as new features and best practices emerge and the project continues to develop (https://kubernetes.io/docs/home/).

While the final tuning and orchestration of these settings will need to be determined by the platform engineering team, we can generally recommend that this tuning should be part of your planning phase. You should reevaluate continuously over the life cycle of a platform.

The serviceability of a platform is similar to reliability, except serviceability looks more at what the golden paths are for the users who need to leverage the platform. For a platform to have serviceability, it needs to meet the user's demands. Unlike self-service, which is a user's ability to reasonably accomplish all reasonable actions without reliance on a governing team, serviceability looks holistically at all the user needs and meets them where they are. Reliability could be considered a measurement of serviceability, but the goal of meeting user needs is more all-encompassing than just reliability. Serviceability is at the core of a product mindset for the platform.

Similar to a critical user journey, the golden path maps the more critical use case for a user of a system. In general, this is the series of steps the user can take and expect the desired outcome.

For a developer of an application leveraging an IDP, the golden path would be something like this:

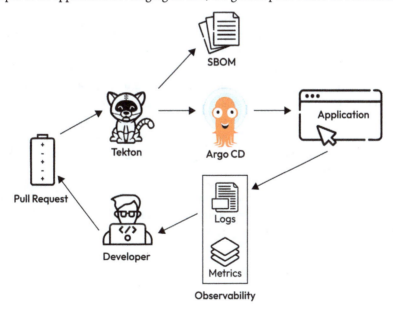

Figure 6.3: An example of a golden path using an IDP

A developer pushes a commit, and then the **Continuous Integration** (**CI**) system does its magic. The application or an upgrade to the application lands in production, and then it sends data back to logging and observability tools for the developer to leverage, enabling them to gain insights into their application's production state and performance.

The conclusion

With this in mind, we can now look at where the SDLC diverges for platform engineering as compared to the development of an application that a platform would run. Along our golden path within the platform, rather than one service or opaque system that does it all, our users interact with a system comprised of different services. Some or all of these services may be homegrown, but it's more likely that these services are a mix of internal tooling and open source technologies working in concert to create the entity known as *the platform*.

Since a platform is built of components, you can also break down your platform's SDLC on a per-component basis. However, as with any system, you must watch out for interdependencies that would require multiple components to be released in concert.

Your developers will need the ability to leverage these systems both transparently and opaquely. This means that while the use of each tool within the platform should be baked into an automated workflow for them, they should be able to see what's happening and debug issues within the individual pieces as necessary. For example, if a Tekton CI job fails, the application developer in our example needs to be able to see the failed job and gain insights into what failed within the job, enabling them to either fix the application or the CI job.

Understanding how to effectively manage the life cycle of a platform while minimizing user impact can be a case of drinking your own champagne, meaning the platform team uses the same processes and technologies they make available to users. The building and life cycle management of a platform can leverage the same DevOps tools; *star (*) as code* patterns such as *docs as code, infrastructure as code*, and *configuration as code* all factor into the build and release of the platform. However, unlike the software application, for end users of a platform these declarative pieces will have the highest impact on them, especially when compared to something such as a banking application for which all of those DevOps aspects are completely black-boxed for the end users. The identity of the customers greatly impacts the weight of one facet of software engineering over the other and, thus, differentiates between *software and platform development*.

That user-centricity is also paramount to understanding what to prioritize within a platform to make it successful. While the users themselves will inform that to a large degree, as we continue through the chapter, we will highlight aspects that we think are most important for a successful platform.

With this new understanding of the SDLC for your platform, we can start to use that knowledge to contextualize what it means to build a platform as a product.

Reducing cognitive load

In *Chapter 1*, we touched on the importance of *reducing cognitive load*. As technology and systems evolve, they become more complex. The high degree of complexity means that understanding every part of a system is a significantly larger mental burden. The old days of monolithic applications on the **Linux, Apache, MySQL, PHP (LAMP)** stack are behind us. Instead, microservices, the cloud, virtual networks, and so on have become part and parcel of the day-to-day operations of a software application.

With modern software architectures, a full stack software engineer no longer needs to understand a server or operating system that application code runs on. The platform takes care of that for them. However, as much as the platform can abstract a lot of the details away, a developer still needs to be able to have awareness and insights into the underlying technologies in so far as they are relevant to their application. Therefore, the platform must strike the balance by making it easy for the users to get from it what they need, without the noise of what they don't.

If we compare and contrast the priorities of the personas, we can contextualize the scope of the platform as it applies to those users.

Platform team	Both	Development team
Tenant apps	Platform availability	Application availability
User management	Team RBAC	Application SLO/SLI
User authentication	Security	Pass/fail CI tests
Networking	Compliance	CD job success
Platform SLO/SLI	Log aggregation	Application upgrades
Platform upgrades	Error and exception tracking	Application resource benchmarking
Platform scaling	Policy violation	
Platform architecture	Quota consumption	
Platform serviceability	Application performance	

Table 6.1: Comparing and contrasting developer versus platform team priorities

As you can see, even with a platform team taking care of so many aspects of how an application is delivered to an end user, the development team for the application still has many considerations. The platform can take care of shared considerations, such as security and compliance, but it's not the job of the platform to guarantee that the app the development team produces is performant. You could consider your IDP a managed service, which wouldn't be a bad definition for the product.

As a managed service seeks to strike the balance between doing and enabling, the preceding table becomes the map to building a successful platform. For example, by investing in tooling around observability and reliability, a platform can support the development team in managing their considerations, while abstracting away everything else it can. A developer who doesn't have to care how aggregate logs are stored but knows how to access them if needed can simplify day-to-day workflows significantly, allowing more cycles to be spent on feature development instead of operational overhead. This reduction of cognitive load can lead to fewer context shifts for the team and, therefore, fewer mistakes and less stress. This way, an investment in a platform is an investment in the productivity and health of the development team.

If we think back to our fictitious company, Financial One ACME, we can imagine how much more the cognitive load matters as it undergoes a cloud transformation strategy. Since the company is not building everything greenfield, its developers have to maintain legacy systems and architectures while also refactoring for the new era. A platform in support of this effort allows the developers to learn less. They just need to know what the platform expects from them as they approach their refactorization and migration strategy.

A platform's commitment to the reduction of cognitive load is essentially the promise to developers that you will help them to work more efficiently. Fewer context shifts, expedient and direct feedback loops, and ease of use all contribute to developer happiness. Happy developers are less stressed co-workers and contribute to a more positive work environment, making it a win-win for the development and platform teams.

Utilizing a platform while balancing cognitive load

While so much of a platform is serving to build an application, it must equally serve to operate the application as well. While the platform's requirements should be well documented and easily understood, humans unfortunately are fairly error-prone; as such, it's incumbent on the platform to enforce its norms. Kubernetes fortunately supports this natively to some degree, with admission controllers and **Role-Based Access Control** (**RBAC**), but you can also leverage tooling such as policy engines to ensure that workloads or actions that do not meet those norms are rejected and an error is given to the user. The sooner that feedback reaches the user, the lower their cognitive load, as they do not have to context-shift to respond to the platform.

So, how are you supposed to reduce the cognitive load for your engineering teams with an internal developer platform?

If you don't know where to begin, start with empathy as you consider implementation details.

Suppose you are one of the engineers using the platform. It is 4 a.m., you are panicking, and nobody else is there to help. Ask yourself, "*Will this be easy to use?*" When you are satisfied with your approach, ask an engineer to validate your assumptions.

Every detail should align with this principle. If core functionality is simple and makes life easier, users will not have to take on cognitive load. As you account for other aspects such as integrations, follow the same approach. Straightforward architectures are often the most effective. If "*Why is this architecture here?*" cannot be answered with "*Because it helps the user*," perhaps it should not be there at all.

There are still significant gains that can be made by fine-tuning the operations of the IDP. Additionally, by setting a bar for the applications that will run on the platform, your team can define a maturity model for them. That maturity model will help define what an application needs to be a successful member of the platform environment, which means that developers can work against it as a checklist, instead of needing to experiment, ask a human, or guess.

Pre-production versus production

A platform covers both pre-production and production for a software application. The part of the platform where the application lands is referred to as the **application landing zone**.

While a platform must consider both, the two should never mix, so the platform must enforce logical segmentations between the environments. It can do this via architecture or a policy. However, architecture is the best and most secure model, which we'll cover further in *Chapter 7*. In the current chapter, while we may make references to security best practices and segmentation strategies, the in-depth explanations can be found in *Chapter 7*. The important takeaway for these environments in this chapter should be how they interact with cognitive load and how the platform aims to reduce that load across both scopes.

Authentication and tenancy

As your platform expands and gains adoption, it essentially becomes multi-tenant. This is due to the importance of ensuring a least privilege policy in a secure environment, as well as helping to ensure users don't trip over each other. With multi-tenancy, it's important to still give the feeling of a single tenancy, which means the existence of the other users and tenants must be hidden from each user.

One of the first integrations you'll add to the platform relates to authentication and user management. Most **OpenID Connect** (**OIDC**) providers can integrate into the common IDP toolchain, so selecting the correct one won't be very difficult. From there, user management is pretty straightforward, although it differs between the common tools. For cognitive load, less is more, so one way for your users to authenticate across all aspects of a platform means fewer login workflows to remember, fewer logins per day, fewer passwords, and a homogenized experience with the rest of the tools in a company.

Let's look at this in a more realistic scenario. Financial One ACME is a bank. This means they're in a highly regulated industry with a lot of very sensitive data to protect. Controlling who has access to what data and how they access it is one of the most important aspects of the company's security and compliance story. It's too much to expect users to remember to exercise all of those security and compliance practices perfectly every time. Thus, the platform needs to enforce that programmatically.

Meeting those requirements in any industry, highly regulated or not, all comes down to RBAC – how you isolate users from each other and guarantee that they have the actions they need without the access they do not. Take a look at the following figure to see an example of a multi-tenant IDP.

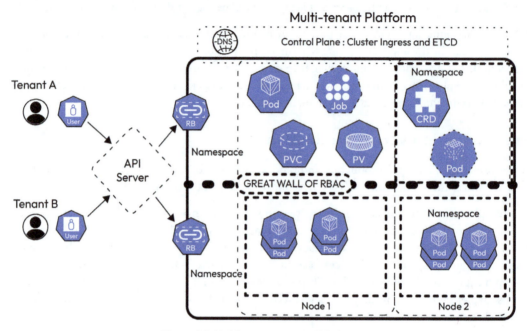

Figure 6.4: Multi-tenancy on an IDP cluster

In this example, workloads from both tenants can share space on a node, but through RBAC, the tenants cannot see the other workloads on the same nodes where their workloads run. Another option would be to also ensure only one tenant's workload can run on each node. However, that level of isolation can have negative trade-offs, as it could impact the **high availability** of those workloads, or require more nodes in order for tenants to have full isolation and high availability. We discussed high availability for a platform in *Chapter 3*, and those same principles apply to an application that your development team is working on. Just as with the platform, high availability is a characteristic of the resilience of an application. This might mean its ability to handle load through scaling, or just its general uptime. Things that impact high availability include location. While you might have three Pods in a **ReplicaSet** for an application, if all three of those Pods are on the same node and it is restarted (such as with upgrades to the platform), then the application goes down; therefore, it is not highly available. Typically, a highly available application will have its replicasets spread across multiple nodes, and if those nodes are all in different availability zones, then the application has another layer of resilience added to it. Developers following best practices for cloud-native platforms will typically expect their application to be highly available, which means at least two other replicasets running on different nodes.

RBAC

Regardless of the tenancy model, the kubernetes system RBAC will become the spine of self-service for a platform. You'll use RBAC to restrict user access to the IDP for only the namespaces and environments where their workloads will land within the IDP. You'll additionally have similar RBAC policies in the end user-facing environment where the application will run. The RBAC policies in the production environment will be less permissive than in the development environments, as it will be necessary to maintain a higher degree of scrutiny on those permissions. You'll learn more about this in *Chapter 7*.

> **Further reading**
>
> Kubernetes' official documentation can be found here: `https://kubernetes.io/docs/reference/access-authn-authz/rbac/`.

Borrowing an example from the official documentation, we can see an example of RBAC where a user who is assigned a role would have the capability to read secrets, but they would not be able to edit or delete them:

```
apiVersion: rbac.authorization.k8s.io/v1
kind: ClusterRole
metadata:
 # "namespace" omitted since ClusterRoles are not namespaced
 name: secret-reader
rules:
- apiGroups: [""]
 #
 # at the HTTP level, the name of the resource for accessing Secret
 # objects is "secrets"
 resources: ["secrets"]
 verbs: ["get", "watch", "list"]
```

Similar to this example, by leveraging RBAC, you set your users up to successfully navigate a platform with appropriate guardrails in place. So, how do you determine what a robust RBAC policy needs to have to strike the balance between access, actions, and restrictions? This is where self-service takes center stage. Let's look at some workflows that your users may expect to have on the platform.

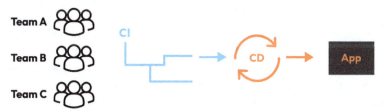

Figure 6.5: Example workflows for IDP users

In the example illustrated in *Figure 6.5*, we have multiple teams and, therefore, likely multiple users within the teams, reliant on the same systems within a platform. Sometimes, they will be reliant on the same systems at the same time. Depending on your CI/CD system of choice, best practices can vary; however, in general, the idea is that the platform team will ensure that the components required for the IDP are capable of scaling to users.

The authentication system used for the Kubernetes cluster can be used for pre-production tools such as Argo CD as well, allowing users to leverage independent projects within the same Argo CD instance.

While it's not a best practice, it is technically possible to do the same access pattern for production-level access; however, when production is being accessed, it's a best practice to have a much stricter set of authentications and an even more limited permission set once authenticated. We'll cover this further in *Chapter 7, Building a Secure Platform*.

Noisy neighbor prevention

It may not immediately be obvious how a noisy neighbor can impact cognitive load, but the protection from such an event helps to prevent production incidents that can be difficult to troubleshoot and resolve, especially in an environment where multi-tenancy is the standard. It also helps users troubleshoot their applications or assists the platform team in troubleshooting production issues by ruling out scenarios that become less likely as more protections are put in place. Since your developer teams should expect to have limited access to the production side of an IDP, and no access to the metrics and workloads for other teams, all work done to prevent a noisy neighbor helps to ensure that those workloads are less likely to enter each team's cognitive load.

In the context of an IDP, a noisy neighbor means a workload that runs in the same environment as another and hogs resources, whether this is through an abundance of network traffic, CPU, or memory utilization, or other less-than-neighborly behaviors. Much like a city would have ordinances to prevent a neighbor from blocking off a street or playing music way too loud, a platform can enforce norms to ensure that one workload doesn't negatively impact another.

In many real-world scenarios, one Kubernetes cluster is utilized for multiple environments. For organizations with more limited cloud spend budgets, dev, QA, staging, and production environments may all be located on the same cluster. Technically speaking, an entire IDP could be one cluster. While this isn't what we'd consider a best practice, it is an extremely common one, utilized to keep costs down. It's in these scenarios where noisy neighbors are most likely to occur, but it can happen even when production is fully isolated, due to a workload either experiencing a software bug or just a higher than normal resource utilization for valid reasons. Containerized workloads have an advantage in protecting against noisy neighbor scenarios in these shared environments, in that those Pod and container definitions can hold the resource definitions, The encapsulation of the workload is a strength, particularly for resource management. While the vast majority of protections against noisy neighbor scenarios should happen as part of defensive programming for software applications, as well as through accurately benchmarking and identifying theresource need of those applications, a platform has the ability to be hardened as well.

How can you achieve that hardening? Here's how:

- Be mindful of the resources that are shared, even with workload isolation best practices enforced, such as etcd, the API server, and the networking stack

- Take preventative measures against CPU starvation

- Cluster autoscaling – the automatic addition (or removal) of nodes

The care and feeding of etcd

etcd is the key-value store that holds every definition for all the Kubernetes objects on a cluster – Pod definitions, job definitions, StatefulSets, DaemonSets, **CustomResourceDefinitions (CRDs)**, and so on are all stored in etcd. This means it can get quite full as platform usage increases. An etcd project recommends a minimum set of resources allocated to the etcd Pods in order to guarantee successful production operations. The detailed recommendations can be found in the etcd docs (`https://etcd.io/docs/v3.6/op-guide/hardware/`) but the functional areas are disks, networking, CPU, and memory. As your cluster utilization grows, so too does the needs of etcd. Generally, the best practices for etcd include ensuring it has priority for CPU, networking, and dedicated disks, with high throughput and low latency. A **solid-state drive (SSD)** should be used for the etcd storage if possible. The etcd documentation has several suggestions for the tuning of etcd, so we won't reiterate them here. It's important to keep up with the latest recommendations from the project itself, as the technology will continue to evolve over time. Since etcd is a critical component to the successful operations of a Kubernetes cluster, watching it carefully with observability and proactively responding to evidence of disk pressure, or other resource issues, will be mission-critical to guarantee platform availability.

Even after your etcd configuration is optimized, the actual usage or over usage of etcd can still result in noisy neighbors. The cloud-native CD system Argo CD, for example, stores its jobs as CRD entries, not as Kubernetes jobs. As the number of teams and deployments from Argo CD grows, so too does the number of entries in etcd. If etcd fills up, the Kubernetes API server goes down. To prevent this, care must be taken to ensure that etcd is healthy and not getting overfull.

One method is to ensure that you're pruning old CRD entries from etcd. In the case of Argo CD, this functionality is built natively into the app. In the case of standard Kubernetes jobs, a time-to-live mechanism can be utilized by applying the time desired, in sections, to the `spec` field of the job definition. With this, after the given amount of time has passed since the job has succeeded or failed, the garbage collection mechanism will run and remove the entry, thus managing the size and health of etcd.

Similarly, other Kubernetes kinds have their own way of invoking a clean-up mechanism. For Deployments, there is a spec for `revisionHistoryLimit`, which will determine how many old versions of the Deployment will be stored. If the number is 0, then in the event of a production issue, the deployment cannot be rolled back, but etcd is clean as a whistle.

Exactly how to tune these clean-up measures will be dependent on the number of users and the number of etcd entries that those users generate. The size of etcd will also be a factor, and that can be scaled by increasing the size of the control plane nodes. So, when to scale, when to prune, and when to consider adding a cluster to the IDP environment will be an exercise in cost management, ROI, and a fundamental part of the platform's life cycle. It will also need to be a factor in the integrations you select for your IDP, but we'll cover cost management more in *Chapter 8*.

Understanding the CPU and the scheduler to avoid starvation

The next resource that is shared at the cluster scope, regardless of tenancy models, is the CPU. As previously mentioned, when we discussed reliability, in Kubernetes, there is a maximum amount of memory that can be used on a node object. All of the Pods and their memory requests are totaled, and if the total memory exceeds what is allocated to the node, a Pod will either be scheduled to a different node with available room or fail to be scheduled. If a Pod tries to use more memory than the node has available, it will be terminated and rescheduled. There's a strict boundary on memory allocation along the nodes. A CPU, at a glance, appears to have the same functionality, but in reality, all CPUs are available to all nodes. This means that while app performance may be primarily a consideration of the development team, the developers shouldn't have to worry about whether the platform can support the needs of an application. While the developer will need to inform the platform of the workload's needs, the platform will need to be able to support it.

The actual implementation for CPU management in Kubernetes is the **Completely Fair Scheduler** (**CFS**), which is the same one that the **Linux kernel** uses. Functionally, what this means is that workloads on a node can use more of a CPU than the node is allocated if the CPU is available.

While this capability to overcommit CPU can be useful, as it can technically support bin packing, allowing a platform to support more workloads as long as they're not running simultaneously, and without needing to reserve more virtual hardware (or actual hardware if you're running on bare metal), there are reasons why you might want to prevent some or all of the CPU from being utilized in this fashion, even if it's technically available.

For example, in a production environment, there may be certain workloads that must always be performant or are particularly sensitive to CPU caching, while other operations could be a little slower or reschedule to new nodes more freely. For these situations, you can better guarantee CPU availability by altering the CPU manager policy.

To understand the full implementation of CPU management in Kubernetes, read the documentation here: `https://kubernetes.io/docs/tasks/administer-cluster/cpu-management-policies/`.

The platform will need to enable CPU management policies in the kubelet configuration in order for CPUs to be excluded from the pool. You will have to define how much CPU to reserve, but you can reserve all of the CPU, minus what is needed to run the Kubernetes-critical workloads. In order to get the desired outcome, where a Pod is guaranteed the CPU it needs at the expense of other Pods, the developer will need to specify both requests and limits as whole integers in the Pod spec, and those two values must match.

Here's an example Pod spec:

```
spec:
 containers:
 - name: myCPUPrivilegedPod
   image: docker-registry.nginx.com/nap-dos/app_protect_dos_arb:1.1.1
   resources:
     limits:
       memory: "200Mi"
       cpu: "2"
     requests:
       memory: "100Mi"
       cpu: "2"
```

Once your node is configured in this way, it is known as *guaranteed* by the platform and, therefore, is able to use exclusive CPU. The **kube-scheduler** (**scheduler** for short) will prioritize it being on nodes that support the CPU needs of the Pod. If that's only one node, then the scheduler will prioritize that; however, if you have multiple or all nodes configured to use exclusive CPU, then although the priority of the Pod will be high and, therefore, less likely to be moved (except in more extreme cases), it could still be placed on a new node.

To ensure that a specific node is always selected, or a specific type of node is always selected, you can label nodes. Doing so helps the scheduler to match Pods to nodes with labels defined in the Pod spec. This further ensures the correct Pod placement, which has use cases beyond noisy neighbor prevention as well.

If you've decided to follow along with a local kind cluster and haven't already, let's set up the cluster now. First, clone the GitHub repository for the book and change the directory to the one for *Chapter 6*:

```
git clone https://github.com/PacktPublishing/Platform-
Engineering-for-Architects.git
//asuming Linux
cd Platform-Engineering-for-Architects/Chapter06
kind create cluster --config kind-config.yaml --name platform
```

After your cluster-named platform is created, set the context to the kind cluster for the chapter. This removes the need to add --context kind-platform to the end of each kubectl command:

```
kubectl config set current-context kind-platform
```

Now that we've made our lives a little easier, we can begin the Pod labeling demo.

Here's the node label command and example output:

```
$ kubectl label nodes platform-worker2 reserved=reserved
node/platform-worker2 labeled
```

Run this command to see the label on the node:

```
$ kubectl get nodes platform-worker2 --show-labels
NAME                STATUS   ROLES    AGE     VERSION    LABELS
platform-worker2    Ready    <none>   3d23h   v1.29.2    beta.
kubernetes.io/arch=amd64,beta.kubernetes.io/os=linux,kubernetes.io/
arch=amd64,kubernetes.io/hostname=platform-worker2,kubernetes.io/
os=linux,reserved=true
```

Our final Pod will look something like this:

```
apiVersion: v1
kind: Pod
metadata:
  name: my-app-nginx
  labels:
    env: prod
spec:
  containers:
    - resources:
        limits:
          memory: "200Mi"
          cpu: "2"
        requests:
          memory: "100Mi"
          cpu: "2"
      name: my-app-nginx
      image:  docker-registry.nginx.com/nap-dos/app_protect_dos_
arb:1.1.1
      imagePullPolicy: IfNotPresent
  nodeSelector:
    reserved: reserved
```

> **Important note**
>
> In order for the CPU management policy to be enforced on previous workloads and new workloads, you have to drain the node and restart it.

Outside of the CPU, the scheduler looks at several other different factors to determine how to prioritize or move Pods, including an explicit Pod priority. If Pod priority is not well understood or the tenancy of a platform is not well known, it may not be immediately obvious to an engineer why they should not place a high priority on their application Pods. As far as a developer is concerned their production application is probably the most important workload. However, this can create noisy neighbor situations if not used well. One over-prioritized Pod seems harmless. In practice, it is a constant denial-of-service attack that worsens with scale. If the Pod is assigned to a replicaset and the instances of it grow with the cluster, then there's a growing number of over-prioritized workloads. If this problem spreads to more than one Pod, then the problem compounds. This degrades the platform service, strains underlying hardware, and impairs troubleshooting. Therefore, the platform may want to limit or prevent such assignments. This can be done with the enforcement of quota or admission controllers.

We'll cover quota and admission controllers later in this chapter. However, if Pod priority is something you'd like to use in a platform, here are the general steps necessary to enable it:

1. First, create a priority class. This kind is not scoped to namespaces but is more generally available cluster-wide. Create the class by saving the following content to a YAML file called `priority.yaml`:

    ```
    apiVersion: scheduling.k8s.io/v1
    kind: PriorityClass
    metadata:
      name: critical-priority
    value: 1000000
    preemptionPolicy: Never
    globalDefault: false
    description: "This priority class should be used for platform
    Pods only."
    ```

2. Next, apply the priority class:

    ```
    $ kubectl apply -f priorityclass.yaml
    priorityclass.scheduling.k8s.io/critical-priority created
    ```

3. Now, when we create a Pod, we can select that priority. Save the following as `pod.yaml`:

    ```
    apiVersion: v1
    kind: Pod
    metadata:
      name: my-app-nginx
      labels:
        env: prod
    spec:
      containers:
        - resources:
            limits:
              memory: "200Mi"
              cpu: "2"
            requests:
              memory: "100Mi"
              cpu: "2"
          name: my-app-nginx
          image:  docker-registry.nginx.com/nap-dos/app_protect_dos_
    arb:1.1.1
    ```

```
        imagePullPolicy: IfNotPresent
    nodeSelector:
      reserved: reserved
    priorityClassName: critical-priority
```

4. After saving the Pod, create it with `kubectl create`:

```
$ kubectl create -f pod.yaml
pod/my-app created
```

There are many other tuning options within Kubernetes to ensure that the scheduler makes the decisions you'd prefer. It is also possible to run an additional scheduler profile or to replace the default scheduler entirely with a customer profile that abides by different rules.

This is a pretty advanced operation, as it requires a deep understanding of how the scheduler works and how you need to change it to optimize the platform. However, if you do decide that this is necessary, we'd recommend starting with an additional profile before replacing the default.

> **Important note**
>
> For OpenShift users, while the end result is the same, the process of implementing an exclusive CPU in OpenShift is slightly different and leverages machine sets. Check the latest OpenShift Container Platform docs for up-to-date instructions on this topic for that product.

Rate limiting and network health

Rate limiting is the capping of network traffic or HTTP requests that a service will respond to from a requestor. Each response has a computational cost, so rate limiting acts as a protective measure against endpoints being exploited, either intentionally or through an error.

A common takedown of a system is a **Distributed Denial-of-Service (DDoS)** attack. While such an attack can be initiated by a bad actor, it can also be carried out by accident due to human error or a buggy piece of software. DDoS is the most classic example of a noisy neighbor. Computationally expensive, it can be executed either via sheer number of requests flooding the networking layer, spending compute, or through requests that would result in data returns that are astronomically large or computationally expensive to get.

Rate limits can be implemented to prevent a large number of back-to-back calls from creating a noisy neighbor scenario. This can be handled in the cluster ingress, such as nginx. But there are also discreet Kubernetes features within the API server that can be leveraged as well.

On the API server itself, your team can set API priority and fairness rules to help prevent traffic flooding. This is especially important if the noisy neighbor situation is coming from an internal source, meaning that cluster ingress is not a factor. For example, if a backup job is misconfigured and it's trying to query data from within a cluster to create backups, or push them too often, applying rate limits can help prevent this job from flooding the network layer. By default, basic priority and fairness rules are enabled, although it is possible to alter or disable the rules entirely if desired. However, that's not recommended.

Cluster scaling and other policies

What other policies prevent noisy neighbor situations? Autoscaling is the most obvious to the authors, although there are surely many more. All of the tuning and planning for how to pack workloads onto the platform only goes so far, without the cluster ultimately needing to grow (or shrink) in size. The most classic of all noisy neighbor scenarios is caused simply by a platform too small to meet the demand of its users.

Where the resource constraints are in the platform will determine how the cluster should scale. If additional resources are needed for non-system components that run on worker nodes, then adding another node to the cluster will do the trick. However, if, despite previous efforts, etcd is filling up or the API server encounters constraints, a larger control plane is necessary.

Observability data is the best measure for when and how to scale a cluster. This data can be collected on compute resource utilization by workloads and workload performance. As utilization increases, if thresholds for remaining resources are met or exceeded, a cluster scaling event can be triggered. Your observability can also be used even more creatively. A common technical interview question is a troubleshooting problem. The solution is always that a cron job uses too many resources and causes system crashes. With observability in place, you don't need a human to discover that; you can use data. Significant spikes in resource utilization, either on the Pod or node level, can be captured and alerted on. But if you needed to respond to an alert, then this is also an opportunity for automation.

And finally, if all else fails, you can use an event-driven system to react to the data collected by the observability stack to restart the faulty Pod. This works and should keep the mission-critical components going, but it doesn't replace a human permanently fixing whatever caused the resource utilization to begin with.

In short, maintaining cluster health and appropriate user and workload isolation is key to ensuring that users of a platform can focus on what matters most to them, helping us to lay the foundation for the next aspect of a successful platform – self-service.

Enabling self-service developer portals

We've already stated that a platform is not a platform if self-service isn't a core tenant of its design. Now, it's time to understand what that means and how it pertains to the promised reduction of cognitive load. The platform team can't be on hand to approve everything all the time. As such, it is imperative to put some control into the hands of end users, without failing to save them from an increase in cognitive load. This dichotomy requires striking a careful balance, which must be discussed and negotiated between the stakeholders to be successful. A platform team trying to control everything too tightly prevents the platform from scaling with the users, as it'll always be limited to the size and location of the team. This means it's time to think about how you help your users help themselves. What should a developer or team be able to do without higher levels of approval? What should be restricted? How do you enable sensible self-service, and how do you enforce reasonable restrictions? We've discussed the importance of treating your platform like a product, but functionally, the product type is a service. Any product sold "as a service" comes not just with a guarantee of features but also a guarantee of engagement and experience from the company selling the **Software as a Service (SaaS)** or **Platform as a Service (PaaS)**. Even if you're not selling your PaaS, you should adopt the mindset of service to ensure that users are at the center of your usability considerations.

Simply put, self-service means that a user can accomplish a reasonable action with reasonable ease, without negatively impacting unrelated users or teams. Understanding self-service as it pertains to your platform requires understanding the needs of your users. It's important to meet your users where they are. If your users are of the kind that spend their whole day on a command line, a **command-line interface (CLI)** may be the tool they need most to get the job done.

However, if your IDP is being used for docs as code instead, in addition to code as code, it's probable that the team leveraging it may benefit from a UI instead of a CLI.

How does the end user want to accomplish their goals? How does the end user want to receive feedback from platform jobs? When we approach self-service, these are the questions we need to ask. Generally speaking, the platform team should have parameters that the developers can operate within. For example, you can guarantee that certain workloads only land in certain locations, and you can tie resource (memory and CPU) limits to namespaces within your Kubernetes cluster.

> **Tip**
> Once you begin down this path to fine-tune how workloads land and consume resources, it starts influencing cluster capacity artificially. It's important to only implement this strategy when it becomes truly necessary, or after creating a comprehensive cluster scaling and capacity plan.

Enforcing quota

A user or team should be able to land their application within the application landing zone. However, they should not be able to do this without reasonable restrictions. These restrictions, such as resource consumption, can be limited to a per-namespace basis. This is done by specifying the kind ResourceQuota. You'll make a YAML file and apply it to the namespace. Continue reading to see what it will look like.

First, create a YAML file; let's call it `your-dev-quota.yaml`:

```
apiVersion: v1
kind: ResourceQuota
metadata:
  name: your-dev-quota
spec:
  hard:
```

```
    requests.cpu: "2"
    requests.memory: 1Gi
    limits.cpu: "4"
    limits.memory: 6Gi
persistentvolumeclaims: "5"
services.loadbalancers: "1"
services.nodeports: "0"
```

Save the file, and now apply the YAML to the namespace. Your command will look something like this:

```
kubectl apply -f your-dev-quota.yaml --namespace=your-dev-namespace
```

Now, your developer can schedule any Pods within their namespaces as they need to, but the workloads have to stay within the quota boundaries provided to them. The creation of a quota for namespaces helps to enable multi-tenancy within a platform. Divvying up resources to users in a way that utilizes cluster capacity without over-taxing it allows users to operate independently, but within reasonable boundaries.

Quotas can also be managed with GitOps, thereby streamlining requests to increase quota or alter quotas from users. If the process for a user to adjust their quota is raising a **pull request** (**PR**) and then waiting for it to be approved, this makes it much easier for them to interact with a platform. Ease of use will be a large factor in the platform's success within an organization.

Simple repeatable workflows

We talked about the importance of simple workflows for reducing cognitive load, and truly, this is a topic we could have stuck anywhere in this chapter. However, this aligns most with self-service, as it is a core feature. For a platform to deliver on the promise of self-service, it must be easy to use, and the uses must be easy to remember.

We refer again to our golden path. A developer building an app pushes a commit. The change makes its way through the CI/CD systems until, ultimately, the app install or upgrade lands in the application landing zone. The journey for that PR must be easy to understand, intuitive, and predictable for the development team. If the team needs to make changes to the CI jobs, or the CD logic specific to their application, then they should expect to have easy access to the governing systems to do just that.

Let's bring back our teams for Financial One ACME. We can expand this example one step further and imagine again that the company includes multiple personas or user types. If we consider a development team and a corresponding docs team, what would each team need from the CI system?

It's safe to say both teams need the ability to land their jobs within the system, to understand whether jobs are successful or failed, and to know whether the overall CI pipeline succeeded or failed. However, they may have different needs for interacting with the CI system. A developer or a DevOps engineer may prefer to use a CLI, whereas a docs team or a non-technical team may prefer a **graphical user interface** (**GUI**) to leverage a system.

The open source CI/CD projects Tekton and Argo CD both ship with a GUI. Both systems also can leverage CLI tools designed to interface with the project. For users of the systems, as long as they've been granted correct access, they can choose their own adventure for the actual interactions with these systems and can determine which is easier, keeping a CLI up to date locally or using a GUI occasionally. By being flexible, these tools and your IDP, should you adopt them, enable users to self-service based on their ease.

However, with something such as CI pipelines, which may take significant time to complete, other options such as API calls or Webhook integrations into common tools such as Slack may more easily meet the users where they are. Just as with inputs, the platform should meet users where they are for outputs, as the fewer times your users have to context-shift, the lower their cognitive load and the easier self-service goals are achieved.

> **Important note**
>
> A CI pipeline is typically a series of jobs; some run synchronously, and some run asynchronously to build and validate an application based on recent changes. This could be one PR at a time (CI at its core), but it could also be less frequent, such as daily.

Even highly technical users love a good GUI, and if they're already interfacing with something such as GitHub or GitLab for a PR review, they may not feel like changing paradigms again. So, while the average user may never need to see the internals of the Kubernetes clusters they're interacting with, we did say that the platform should drink its own champagne. This includes best practices such as meeting users where they are. Some Kubernetes solutions such as OpenShift ship with a GUI layer included. For the ones that do not, Lens (`https://k8slens.dev/`) provides an easy-to-use GUI.

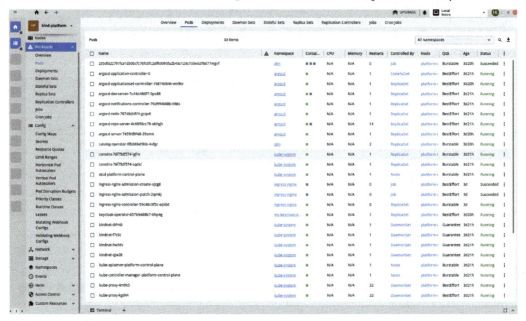

Figure 6.6: The Lens GUI connected to a local kind cluster representing an IDP

To summarize, developer self-service is about meeting your users where they are and giving them the tools to be successful with a platform. Tshis means understanding their needs but also giving them the capabilities to integrate the platform to their way of working, not making their way of working fit the platform.

Landing, expanding, and integrating your IDP

If you followed our recommendations in *Chapter 3*, you've interviewed the users, you've mapped the use cases, and you've designed your platform. Now comes the fun part – turning it into something real.

Let's look at the most basic IDP. What does it do? Most importantly, users can authenticate to it with appropriate access and permissions. It has a CI system, a CD system, at least some basic security tooling such as scanning, and these days, something to generate an SBOM. What about the production location for the software that the developers are writing? Where is that? How does the promotion process look? Let's look again at our IDP golden path from earlier in *Figure 6.3*.

There are a lot of pieces in this drawing, and software development should be iterative, which means we're not going to be landing everything seen here all at once. So, let's look at landing an IDP. What does the thinnest viable platform look like?

It's probably something like this:

- Authentication and RBAC
- Security checks
- CI/CD
- An application landing zone

Technically, a CD system isn't necessary, as long as there's some promotion process for application features and enhancements to reach production. With such a scaled-down approach, or thinnest viable platform, it's fairly easy to see how to land an IDP within an organization. The appeal of a new and shiny product may even drive users to play with it and help drive initial adoption. However, keeping those users engaged and attracting the more reticent will require expanding upon the initial offering and ensuring that what users use is easy to understand and performant.

Expanding your IDP, though – how does that look? Expansion may look something like capacity planning, but how will you scale and size your resources as the adoption of the platform grows? In Kubernetes, this will mean figuring out cluster sizing and/or cluster count.

The expansion will also include features and enhancements, as well as planning not just for the development and rollout but also the operations of those enhancements. What are the impacts on the platform of adding another feature or integration? How do you measure the success of an enhancement, and how do you use data to determine its return on investments?

Enforcement of platform-specific standards

If we refer back to the beginning of the chapter, we mentioned that it's important that platform-specific standards are easy to consume for end users, helping to guarantee a low cognitive load, and that this must include the enforcement of those standards.

There's no one set way to do this, since norms can differ from platform to platform. However, we have found that there are a few things that work well in terms of both process and technology for you to adapt to your organization.

Maturity models

Maturity models fall slightly more into the sociotechnical aspect of cognitive load reduction. Defining maturity models for software applications as they onboard to a platform can help developers know they're building out their application in the right way. This also helps to ensure that the reduction of cognitive load doesn't result in a worse experience for end users, due to something being overlooked or forgotten. Essentially, it ensures that the interests of both parties are aligned from a platform perspective.

An example of a standard for maturity would be that a new application would only onboard to a platform after it has been configured for high availability and disaster recovery, if it has unit tests, or if the team has written and committed their Prometheus queries and Grafana dashboards for application observability. These types of models can be enforced with CI jobs that check for the presence of whatever is deemed appropriate by the platform, or they could simply be published as guidelines for a team to follow.

Expanding a platform with common platform integrations

Taking a platform beyond the thinnest viable platform and into a robust system will require leveraging new integrations and better harnessing the initial feature sets. Some common and useful integrations are outlined in this section.

Static analysis

How does a user know whether they're matching with expected maturity models and platform norms? How do they receive feedback if they are not? One method for establishing this enforcement and feedback loop with users is via static analysis. When the development team submits a PR, then CI jobs kick off. It is during the first CI jobs that a static analysis tool is best positioned; this way, if the PR fails, the check compute is not spent needlessly on running tests or generating SBOMs, and the feedback loop to the end user raises the error much faster. Reducing time to feedback helps to keep developer velocity high and cognitive load low, since developers don't need to remember to check back with the platform after a significant delay.

Example static analysis tools that can be leveraged at PR time are security audits that compare image versions to known **Critical Vulnerability Exploits** (**CVEs**) to ensure that vulnerable code is not being pushed. At the time of writing, **Snyk** is a popular choice for this, as it is open source and has free options. **CodeQL** is another popular choice and is also free for open source projects, so it's very easy to find examples of it in use. Both of these tools are popular because they integrate well with GitHub workflows.

For the enforcement of norms, however, there are also static analysis tools that verify whether an application conforms to cloud-native best practices. From the **StackRox** community, **KubeLinter** is an excellent example of this. It ships with a variety of checks but can be configured to skip checks that are not desired for a platform, or to accept custom checks.

Both the aforementioned static analysis tools leverage GitHub workflows, which are declarative YAML to define what actions should happen within a repository once a PR is raised. For example, to run KubeLinter against the main branch of a repository with GitHub workflows, a `.github/workflows` directory would be added to the GitHub repository, and then a YAML file would be committed to it. That YAML would look something like this `kubelint.yaml` file:

```yaml
name: StaticValidation

# Controls when the workflow will run
on:
  # Triggers the workflow on push or pull request events but only for
the main branch
  push:
    branches: [ main ]
  pull_request:
    branches: [ main ]


  workflow_dispatch:

jobs:
  lint:
    runs-on: ubuntu-latest
      # Checks-out your repository
      - uses: actions/checkout@v4

      - name: Scan yamls
        id: kube-lint-scan
        uses: stackrox/kube-linter-action@v1
        with:
          directory: app-of-apps
          #config: .kube-linter/config.yaml
```

Admission controllers and policy agents

Admission controllers are another feature of Kubernetes. They come in two flavors, validating and mutating Webhooks. Both reduce cognitive load and also enable self-service. However, they're fairly advanced topics and may not be part of your platform MVP, or even in the top five integrations. As the platform itself evolves in maturity, these features can be adopted.

If a mutating Webhook allows a platform to enforce a norm such as an environment variable transparently to the user, this allows them to worry less about the platform itself and more about the application. Then, if the environment variable changes, the developer doesn't need to do anything differently on their side. A validating Webhook checks that the object being applied matches the expected parameters. You can leverage both types against workloads, and generally, you should always use a validation Webhook if you're using a mutating Webhook to ensure that nothing else has incorrectly modified the object after the Webhook has responded.

Admission controllers are powerful but somewhat complex. They require setting up a server to run the controller and the Webhooks and responses to be written. Due to their size and complexity, relative to other methods of enforcing platform norms, they're not suitable for a new IDP, but for large organizations with many users, they can become a powerful tool.

Admission controllers can be bespoke, but open source policy agents such as **Open Policy Agent (OPA)** or **Kyverno** leverage the same features and can help simplify their usage. Leveraging policy-based norms can help platform performance and prevent collisions between workloads.

Observability

To be covered in greater depth in the next section, observability is something that should exist from day one in the IDP and can and will be expanded. Observability should be considered a living integration, meaning that it is always being grown and modified to match the life cycle of a platform.

Some common observability tools are Prometheus, Grafana, and Loki, and they can be used individually but are better configured in conjunction with **OpenTelemetry (Otel)** and the Otel protocol. When you combine these with Thanos for long-term storage, you can give yourself a little more data to analyze as well, which can be nice to understand the context and history of observability data.

Respectively, these tools specialize in metric queries against Kubernetes objects as well as analysis of those metrics, visualization of metrics, and log aggregation. Additionally, this toolset can be used by single tenants with multiple users, or as a multi-tenant tool. The user authentication and RBAC for the IDP can be leveraged for these, just as they could with the CI/CD solutions we discussed previously.

The MVP of a platform and its golden path should be the main focus of the initial observability implementation. Observability should be done in a way to measure user satisfaction, and **Service-Level Objectives (SLOs)** directly maps to a user's success with the platform. Therefore, as a best practice, make sure that the core components of the golden path are available, have a set target for success and latency, and ensure that the platform team can leverage observability to make data-informed decisions, including tracking usage of certain features and the failure rates within.

As an example, let's say that Financial One ACME initially shipped a new platform MVP, with Tekton included for CI and Argo CD for CD. But thanks to the observability stack, the platform team can see that only 1% of users leverage the Tekton integration. This is an indicator that the CI needs are being met elsewhere. Thanks to this data, the team can take the decision to deprecate the Tekton deployment and reclaim some cluster capacity, while letting users continue to work in their preferred way. Alternatively, they could work to understand why the feature of the platform isn't being used and seek to help users adopt the new CI system.

Integrating your platform

Integration of the platform doesn't just mean integrating components with each other; it also means integrating the IDP into a company's way of working. This entails a service-first culture, user-centricity, feedback loops for operational excellence, and meeting users where they are.

Once again, we will go back to Financial One ACME. This company needs to balance both a new-age technology stack and a legacy stack. An opportunity exists here to meet these users where they are by integrating the existing workflows for authentication into the new platform. If they're already using GitHub or GitLab, they're likely using some kind of single sign-on, which should be compatible with any Kubernetes-based IDP.

Integration into a company's way of working may mean certain options to solve some problems don't meet the users' needs at this time. For example, if the reason for low Tekton adoption in the Financial One ACME development team was because GitHub workflows with Argo CD are capable of handling all of their CI/CD needs, then that integration with Tekton may be an effort with a low return. Developers who are refactoring a monolith may find more benefit from being able to use a shared CI tool. This would help them to create a common set of tests, and checks between the old monolithic system and new system, ensuring that they're maintaining parity during the transition. In the example of low Tekton adoption, an upfront analysis of the decision may prevent wasted time, instead that time may be better spent on the integration of a data transformation tool such as Apache Airflow or Argo Workflows.

Since successful IDP adoption is going to be data-driven, a key aspect of this will be **observability**. The observability of a platform will help to drive its value to consumers and the team that must maintain it.

Architectural considerations for observability in a platform

Self-service and cognitive load – we've already alluded to the importance of observability in these topics. Your platform cannot act in service to its users without observability prioritized. In other words, you should expect to watch and measure everything.

Observability comes in two flavors; the first is observability for the benefit of the platform team. This is the way the platform team will collect data and information to help build their reliability. Using the site reliability practices of SLOs, the platform team can measure customer satisfaction by setting SLOs for the platform and then creating observability in support of them.

The good news is that, in the Kubernetes world, this is easier because of objects and microservices. The bad news is that there are a lot more things to measure than ever before due to microservices, and it's hard to know whether you've captured everything. Your observability should have a life cycle, like your software. Start with good, iterate to better, and realize that observability is a living thing that will likely never reach a completed state.

There's a standard observability toolbox in Prometheus, Grafana, Loki, and other **Free Open Source Software** (**FOSS**) projects that help to observe and measure workloads and provide insights into their operations. Often, the consumers of observability data are **site reliability engineers** (**SREs**) or DevOps engineers, but any user who is concerned about the operational excellence of their application should be able to easily review and use this information.

Equally important observability tooling that is more applicable for troubleshooting a production issue of some kind includes networking tracing, liveness probes, and stack traces. A platform should supply the implementation and validation of this toolchain and make it available to users in a restricted fashion, ensuring security and compliance are upheld.

Naturally, the log and telemetry data retention policies introduce their own scaling concerns, and logging and backup systems can be the worst perpetrators of noisy neighbor behaviors. Thus, the care and operations of these systems fall to the platform team to ensure a smooth integration into the pre-production and production environments. Other best practices around observability, such as sanitizing logs and keeping platform and application logs quarantined from each other, are important, but we will cover these topics in more detail in *Chapter 7* when building a secure platform.

Observability in a platform

Observability is, to put it simply, the way you measure your platform. In a service-based approach to a platform product, the SRE practice of an SLO or multiple SLOs is designed to uphold the golden path. If you think about your SDLC, that path should have been previously defined in the planning phase. An SLO that supports a user to ensure their success and happiness as they journey along a golden path represents a best practice for an observability strategy.

If we refer back to Financial One ACME, as well as the golden paths that we've defined for a successful IDP within this organization, we can see an observability strategy start to develop. The platform team can measure the Pod health for the most critical items in a workflow. If Argo CD is crash-looping or a component Pod is not ready, then you know that user expectations around CD are not being met.

If jobs are failing for security checks or SBOM generation, then the security and compliance systems need attention. This could block PRs from being merged and interrupt developer workflows as they try to determine why necessary checks aren't succeeding. All along the golden path there is something we can measure that will help inform platform health and developer satisfaction.

Observability in a platform should be in support of SLOs. These measure customer satisfaction through the collection and analysis of data. When we discussed data-driven decision-making for platform evolution in *Chapter 1*, we espoused the value of it as a way to maintain the product mindset for a platform. SLOs are a concrete, data-driven way to create a feedback loop that your team can iterate against.

If your organization cannot meet an SLO, then users are probably not happy. However, even if you're meeting your SLO target, SLOs can still be improved. It's reasonable to start with an SLO such as *"jobs will succeed 80% of the time"* and increase that number to 99.99%. This way, you can still evolve a platform without increasing the cognitive load on users.

> **Important note**
>
> 100% is a bad SLO target because it's an impossible metric; however, the concepts of three, four, or even five nines of availability can be applied.
>
> For further reading on SLOs, we recommend *Site Reliability Engineering: How Google Runs Production Systems.*

Centralized observability – when and why you need it

What happens when a platform hits a critical failure? How do developers, DevOps teams, SREs, or whoever is responsible for responding to production outages figure out what's happening? Immediately logging into the production system and debugging and troubleshooting is one option, but it's not the most secure one. Ultimately, it may become necessary, but it should be considered a last resort, not the first step.

How do you know that your observability stack isn't the man down? In a single-cluster IDP, these questions are easy to answer. But in a multi-cluster or multi-cloud IDP, those questions become more complicated. This is where a centralized observability stack becomes critical for the success of any application, including the IDP itself. A centralized observability stack allows for environment-agnostic observability. The centralized system can reach out to the IDP clusters and validate their liveness with probes, or it can simply alert on the absence of data. If the IDP clusters stop calling home or shipping logs, that's a good indicator that there's a problem.

Important metrics

Since a platform can be prescriptive about how and where workloads run, it can also be prescriptive about how workloads are measured. For example, the DORA DevOps metrics can be collected and used to give indicators of quality of service or service health. A platform can support this by making services available to expose the necessary data points, or it can enforce the calculation of the metrics by forcibly collecting the necessary metrics, since it's the controlling entity of the environment.

While SLOs and service-level indicators cannot be fully standardized across multiple services, a platform should seek to support the collection, aggregation, and retention policies of any critical metrics for an application. Those metrics must then be made available easily and with proper RBAC. This allows users to self-service defining metrics and troubleshooting issues with the software.

Observability in service for developers

Observability for the consumers of a platform looks a little different from what the platform team might see and do. While the toolbox is roughly the same, the data needed is a little different. In addition to their application performance and uptime, developers on the platform will care more about failed CI jobs, DORA DevOps metrics, application logs, and exceptions. They need the ability to see all of their observability, with no platform-level observability or observability from other tenants. This not only helps maintain a security and compliance posture but also reduces cognitive load, since all the data from all the sources can make it hard for developers to understand what data they need and how to interpret it.

There are several ways to approach achieving this clean signal to developers. The observability offering Thanos is specifically designed to be a long-term storage solution for Prometheus metrics, natively supporting multi-tenancy. The query language is still Prometheus' PROMQL, which creates a familiar way of interacting with observability data in Thamos. It does, however, expect a somewhat significant amount of dedicated cloud storage to support its operations.

When we get into the implications of storing and serving observability data in a platform, the issue of scaling the platform takes a new shape. We'll dive into cost management more concretely in later chapters. However, there are a few high-level considerations we can take a look at now.

As setting up Thanos requires creating object storage and allowing access between it and a platform, it adds another layer of complexity to the platform. The storage must be set up, and the networking paths that connect it to the rest of the platform must also be created and maintained. This pays dividends for large-scale operations that need to keep observability data around for a long time, but it's not ideal for smaller IDPs or early versions of the platform. Conversely, Prometheus can be used the same way, but it can also keep the data in a persistent volume via local mode or not persist the data at all, meaning that if the Prometheus Pod were to restart, the metrics it had known about would be lost.

It's possible for teams within an IDP to have different needs for metric longevity; as such, there is a fairly common pattern of multiple instances of Prometheus using federation.

Federation is essentially one Prometheus instance that scrapes the data it needs from another. These instances can be on the same cluster or across different clusters. In the use case of a multi-tenant IDP, Prometheus federation is a logical choice. The platform owns the application landing zone and, therefore, will hold the source of truth for resource utilization within its observability. However, it's not wise or secure to open that up to all users if the development team also uses Prometheus to measure their application health through things such as canary routes and synthetic probes. The developers may want to tie that data together with the data that the platform owns to get a full picture of their application performance. By giving the developer Prometheus instance federated access to the data it needs, developers can get that information without exceeding their RBAC. Additionally, this model keeps cognitive load low by creating a useful **signal-to-noise** ratio. A signal-to-noise ratio is the total number of signals that a developer receives and the number of signals that are actionable versus noise. The less noise a developer is exposed to, the more effectively they can parse the data they need to review.

The actual implementation for this isn't very difficult; it's a few new lines in the Prometheus YAML file. The important takeaway is that the applications can get information from the platform and any necessary shared services, but not from each other. Since application-level observability data is more likely to have personally identifying information or other sensitive data in it, this helps the platform team guarantee that security and compliance best practices or requirements are followed.

The recommended example for federation in this case would be cross-service federation. Here, the most common model is to have a central Prometheus server that does federation and is scraped by other Prometheus servers that may not support federation.

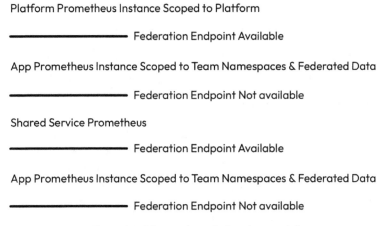

Figure 6.7: A Prometheus federation model

The YAML for the platform Prometheus instance would look a little something like this:

```
global:
  scrape_interval:     15s
  evaluation_interval: 15s
  external_labels:
      primary: 'platform-prometheus'
scrape_configs:
  - job_name: 'prometheus'
    scrape_interval: 5s
    static_configs:
        - targets: ['api.example.com:3000']
  - job_name: 'federation'
```

```
scrape_interval: 60s
honor_labels: true
metrics_path: '/federate'
params:
  'match[]':
    - '{job="github-exporter"}'
    - '{job="namespace-cpu-usage"}'
static_configs:
  - targets:
      - "prometheus.tenant-1-app:9090"
      - "prometheus.tenant-2-app:9090"
```

In the *Figure 6.7* example, there is a shared service, and it may be difficult to understand what kind of service would be important to multiple application observability stacks. One use case would be for DevOps DORA metrics. In DORA there are four key areas that a team would measure if they're looking to implement a DORA maturity model:

- **Mean time to resolve**: How long it takes to recover from a failure
- **Change failure rate**: How often deployments lead to failures
- **Lead time to change**: How long it takes for a change to reach production
- **Deployment frequency**: How often changes are deployed

To get this data, the CI/CD systems would need to be queried, or they can be configured to send data via a Webhook, and using Prometheus to obtain and normalize said data, would allow all compared data to exist in the same format, which makes calculation easier. The implementation details for this would involve a Prometheus exporter. A Prometheus exporter is a service that queries data from systems that may not be immediately Prometheus-friendly but then transforms it so that it is in a compatible data format. For example, to calculate the lead time to change, a Prometheus exporter would get data from the CI system, normalize it, and make it available via either a query to the Prometheus API or federation.

Whatever observability implementation paths you ultimately use, the data and its criticality should be made available to everyone who can leverage it, for the success of the platform and the products it supports. Do this with care and also with an understanding that data is a key asset for users.

Opening your platform for community and collaboration

We often use the term *open source*, but open source doesn't always mean free, and vice versa. However, when we discuss FOSS, the most important aspect of the software is not the technology but the community behind it. Open source communities foster collaboration and help to organically grow software in ways that revolutionize the industry.

When we previously discussed learning about your users, it was largely in the context of getting things off the ground, but you need to maintain that same user-centricity in order to keep the product running and relevant. The best way to harness the power of your users is to bring them to the table with the team, creating a community as seen in FOSS projects that invites collaboration, contribution, and communication. By doing so, you can have a data-driven, user-led platform that evolves with the organization it is intended to serve. Oftentimes, a project lifts off, but the roadmap beyond that is fuzzy. As the adrenaline of the initial push to an IDP fades, the risk of losing momentum and becoming complacent develops. The community aspect should then act as the fuel to keep the vehicle moving.

At this point, you're likely asking, how do we go about creating this community? How do we bring people to the table without undoing the reduction of cognitive load we've been striving for with our platform?

There's no one magic button solution for this but, rather, several things that can be done in concert, such as adding a contributor guide to a repository that sets expectations around what kind of contributions are desired and how contributions will be reviewed.

Another aspect is that some of the tools you might include in your platform can help to entice collaboration. Policy engines, for example, entice collaboration because users can propose new policies or adjustments to their policies. By having a repository for policies that is open to end users, you can invite PRs against the policies for users to propose new permissions or adjustments to existing permissions. As we discussed with self-service, this leverages a known useful workflow of Git PRs and peer review.

Similarly, the use of quotas for an environment is a configuration that can be managed with GitOps; a PR for more quotas and some GitHub workflows can be leveraged to help developers request quotas. In both examples, the leveraging of Git repos and PRs allows users to interact with a platform in a more intuitive way that fits with their workflows. Once again, meeting your "customers" where they are helps to reduce their cognitive load, while enticing collaboration is necessary to ensure that the platform continues to meet their needs.

A platform should drink its own champagne. If the product security team has demands on the application that a platform must support, such as the generations of SBOMs or automated penetration testing jobs, the platform itself should be evaluated against the same security standards. Doing this is not only an exercise in intellectual honesty, ensuring that the way security standards are enforced is reasonable and doesn't interfere with serviceability and self-service, but also a pathway toward engagement from your internal community in the development and hardening of the platform.

A discerning reader may wonder, "*How do you reconcile the dichotomy between the reduction of cognitive load for users and their engagement in the platform experience?*"

This is an important question, as endless requests for feedback may result in silence, since the mental load of generating feedback may be high especially if users don't feel that feedback will be actioned.

There is no single approach to gaining community engagement that is guaranteed to work, but when we look at some best practices from around the FOSS community, we can see there's a trend.

Planning in the open

Regular planning meetings and community meetings that are attended by the platform team, as well as being open to anyone to attend, help to create and foster the open collaboration needed to grow a community culture around a platform. In a Kubernetes project, most of the project meetings are on a public calendar, and there are mailing lists that individuals can sign up for to keep abreast of the latest and greatest news coming from each project. Regular product updates and plans that are emailed out and transparent to a company will help to maintain interest and engagement, hopefully resulting in the proactive feedback necessary for the long-term success of the IDP.

Accepting contributions

Accepting contributions is probably one of the more logistically difficult sociotechnical aspects of landing and expanding an IDP. However, some open source IDP projects lend themselves very nicely to this paradigm. The recent open source darling Backstage provides, among many other features, a plugin-capable architecture. In other words, by laying down Backstage as the IDP framework when a team finds they're missing functionality or would benefit from some additional quality-of-life features, they can leverage the plugin framework to propose the new feature via a PR to the internal IDP project.

While you don't have to use this exact framework to achieve the same results, the pattern that Backstage exemplifies allows for easy contributions from users and a standardized way of proposing changes via hands on contribution, instead of just sending a request to the platform team's backlog.

Summary

In this chapter, we looked at how to build for developers and the importance of cognitive load. Protecting users from undesired outcomes positively influences their cognitive load and gives them fewer concerns, and fewer things to debug if something goes wrong can help keep developers working efficiently. Additionally, the relationship between cognitive load and self-service was explored, and we took a look at some common tools and patterns to deliver a platform that meets the users where they are. If you were following along with a kind cluster, you may have even gained some hands-on experience with these recommendations. While we touched briefly on the security aspects of each of these topics, we'll cover each of them and more topics in depth in the next chapter.

Part 3 –
Platforms as a Product
Best Practices

In the last part, we will provide you with the tools to optimize your platform for cost efficiency, enabling your users to do the same. We will achieve this by outlining the simple steps required to establish transparency within your cost landscape and by providing you with best practices to reduce your infrastructure expenses. From infrastructure costs, we will move on to technical debts, which can negatively impact the maintenance costs of the selected technology stack if they are not dealt with correctly. You will learn about tools and frameworks to evaluate technical debts and the importance of documenting what decisions you make. Finally, we will take a look into the future. You will learn about the imperative of change and why you have to become an active part of driving change. You might discover a different perspective on your golden path and some ideas about future relevant technologies.

This part has the following chapters:

- *Chapter 7, Building Secure and Compliant Products*
- *Chapter 8, Cost Management and Best Practices*
- *Chapter 9, Choosing Technical Debt to Unbreak Platforms*
- *Chapter 10, Crafting Platform Products for the Future*

7

Building Secure and Compliant Products

In the digital age, cyber crimes are on the rise. While not every organization will have the hard compliance requirements of a bank or a government agency, the security standards and best practices for those highly regulated environments can and should be generalized to your platform. Security at every layer of an organization helps to prevent a security breach down the line.

By the end of this chapter, you should have gained a better understanding of security standards, frameworks, and trends. This includes tips for understanding and leveraging a **Software Bill of Materials** (**SBOM**), understanding open source projects for platform security, and understanding policy engine technologies (with examples and use cases). You should be able to use these learnings to define the right actions to secure your platform without limiting your capabilities and ensure the app delivery process will provide hardened and secure software/container packages.

As such, we will cover the following main topics in the chapter:

- Reconciling security to the left and Zero Trust
- Understanding platform security – how to build a secure yet flexible and open system
- Looking at SBOM practices
- Understanding pipeline security – what you have to consider to secure your **continuous integration/continuous delivery** (**CI/CD**) pipelines
- Understanding application security – setting and enforcing policies
- **Free and open source software** (**FOSS**) for platform security and how to use it

Reconciling security to the left and Zero Trust

Security to the left and **Zero Trust** are the buzzwords du jour in cybersecurity. These buzzwords – or buzz phrases if you will–most certainly will fade into obscurity, but the practices they represent will continue to be best practices for years to come.

Security to the left looks at the process of building and delivering software as a linear flowchart that reads left to right. That chart would look something like this:

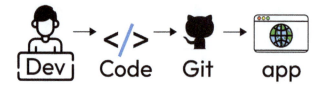

Figure 7.1: Simple app development workflow

In this very simplified example, a developer writes code that is then placed in source control and ultimately comes out as an application for consumption by users. Looking at this security workflow on the right side, at the application layer itself, is important but too late. There are already three other obvious places where a lack of security can create vulnerabilities that could be exploited.

Solving security at the personnel level is the essence of security to the left, but that's not where security ends; it's just where it begins. Security must be centered at every step along the flowchart so that when we get to more realistic examples, we can see how security needs to expand with the footprint:

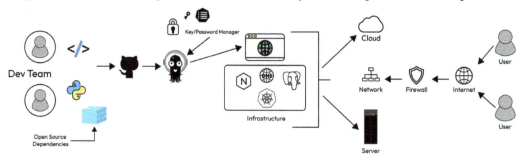

Figure 7.2: Expanded development and delivery workflow

In the preceding diagram, you can already see some common security best practices implemented at the point where users are trying to interact with the application and its supporting infrastructure. However, security at the dev team and open source dependencies' levels, within source control, with CI/CD, and on the application, itself is not addressed. Even storage for keys and secrets, while representing a best practice, also needs security applied to the accessing of those passwords. Security to the left helps you develop your security story from the beginning of the product life cycle through to the delivery of the completed application to the end users. In platform engineering, this can create a feeling of dissonance between the most secure platform and the platform that achieves perfect self-service. Since the platform needs to support developer self-service, then the platform owning the full security story starts to run counter to the flexibility that accompanies self-service. It can, therefore, impose enough restrictions for it to feel like a high barrier to entry and thus jeopardize the adoption of the platform and developer happiness.

In days past, *Trust but Verify* would have been the security model to solve this. The meaning is self-explanatory. You trust the developers have done everything necessary to maintain the desired security posture for the application, but at the platform team, you don't own the security posture end to end. That platform would verify to the best of its ability that all the right things have been done without intruding on self-service.

These days, security best practices have shifted, and *Zero Trust* is the name of the game. Zero Trust essentially assumes everyone is a bad actor (whether or not that's intentional is irrelevant). To maintain this security posture, the platform needs to conform to best practices, but it cannot take on the responsibility of the application. In other words, the platform needs to provide all the scaffolding to the dev team and the stakeholders to support a secure and compliant product. For example, if the Python language is needed, then a secured Python binary, either internally secured or from a trusted vendor, could be supplied within an image registry accessible to all users and applications of the platform. Using Python-Slim from the Docker registry is also a more secure choice and more readily available. The Slim image is probably fine for most use cases. From there, a reasonable and self-service attentive restriction would be to deny workloads that don't use images from a known safe source. A check in the CI pipeline can handle this. Pushing the check as far left as possible saves everyone time and also avoids spending compute processing a change that isn't conformant to the security posture. However, adding a check in that part can be a little painful as it involves writing a job to scan, analyze, and then reach a decision based on the contents of all Dockerfiles in the repository. Those could be nested within sub-directories, and while it's not the most impossible challenge, it can be painful. Additionally, writing this type of functionality into CI pipelines should be considered out of the scope of the average platform team.

From the platform perspective, an alternative approach would be to use policy engines and admission webhooks to reject any Pod definition that doesn't leverage a trusted image source. This isn't as far left as the ideal, and hopefully, the development team and the processes they follow prevent cases where this would be necessary, but in Zero Trust environments, this policy would act as a last fail-safe to ensure only the correct software is being promoted to production environments. It could be deemed reasonable that images from more public sources are acceptable for prototyping within the IDP, and therefore, the fail-safe is only necessary for production environments. This allows the platform to continue to act in service to the team and not unnecessarily impede them.

Another example would be to only accept commits into a GitHub repository if the commit is signed. The signature identifies that the author and the code hasn't been tampered with since the signature was applied. This can be enforced via webhooks on a GitHub repository, although how much influence the platform team has over GitHub organizations for any company may be limited; the security team (if present) will likely require this. And while it's a good example of Zero Trust and security to the left working together, it's again most likely out of the scope of the platform team.

However much these phrases may sound like empty aphorisms, teams fail if they ignore the fundamental principles they represent. Security done right is one of the best investments a company can make. Shifting security left means testing it early and testing it often.

Every major security breach and vulnerability could have been caught with testing. There are various types of testing, including some specifically executed by security professionals, but regardless of whether it's penetration testing or just basic negative testing in your quality engineering process, security and compliance should be tested regularly to guarantee there are no unexpected surface areas. This could look like using software in unexpected ways, or simply validating that the permissions settings for roles within the organization are correct.

Now that we've introduced the concepts of security to the left and Trust but Verify, let's look at how to build a system that is both secure and flexible.

Understanding platform security – how to build a secure yet flexible and open system

The platform isn't the totality of an organization's security posture; rather, it's part of an equation. When assessing how to integrate cybersecurity or DevSecOps into the platform, a balance must be stuck. Pushing security to the left helps to reduce the efforts the platform team needs to exert, but a clear and defined scope helps everyone to understand their part of the security story.

Breaking down the problem into consumable chunks

Security and flexibility can also feel like two words that stand diametrically opposed. Good security is inherently inflexible; however, it's possible and necessary for an IDP's success to balance both. How do we achieve this? Step one is **scope security**.

The first part of scoping is to understand what the minimum level of security is that's required. Obviously, we should always do more than the bare minimum, so if you want to understand what a maximum level would look like, it's not the worst idea, but it may artificially increase the scope and cause you to be unable to see the forest for the trees. Therefore, it's our recommendation to start with a narrow focus. Once you know what the minimum level of security is that the platform must enforce, then you start looking at what the minimum level of security is that the platform must support.

When you go down the path of security, it's very easy to quickly realize how dangerous the world of the internet is and overcorrect. These overcorrections can increase cognitive load and create a barrier to entry for using a platform. If a good developer is lazy, then the platform will need to help them to be lazy, not introduce additional layers of complexity. It's for this reason that while the platform cannot own the entire cybersecurity posture of an organization, it plays a pivotal role in said posture. Security is good; security theater is not.

Knowing how much is too much and how much is just right is a skill that gets honed over time, and there's nothing to cleanly say when a security measure is too much or just right. The answer will always be: *It depends*. For example, securing an environment so that it is not accessible to the public internet may impede the ability of people working on a project to work from home, but if that project involves space shuttles or nuclear reactors, then the air-gapped environment is correct, and not an overcorrection.

Many security experts have spent years researching security and defining solid definitions of what it means to be secure and compliant. The **National Institute of Standards and Technology** (**NIST**) division of the US Department of Commerce is built up of such experts who regularly publish new standards and update pre-existing standards as the industry evolves.

It's worth understanding the work of these organizations to help guide you in developing your understanding of how security and a flexible **Internal Developer Platform** (**IDP**) intersect. Since these agencies typically publish standards for companies that work with the government, it's worth noting that their publications are intended for a specific type of audience and do not replace talking to a cybersecurity expert with experience in your specific industry.

Addressing the OWASP 10

If you're still not sure where to start, another organization has published guidelines that make a great starting point for scoping security. The **Open Worldwide Application Security Project** (**OWASP**) (owasp.org) is a non-profit foundation focused on cybersecurity. As a respected group of security experts, their *Top Ten* list has become a guiding beacon for anticipating and preventing security issues within software. Every so often, they re-publish this list, and as of the 2021 publication, this is their current list:

- *A01:2021–Broken Access Control*
- *A02:2021–Cryptographic Failures*
- *A03:2021–Injection*
- *A04:2021–Insecure Design*
- *A05:2021–Security Misconfiguration*
- *A06:2021–Vulnerable and Outdated Components*
- *A07:2021–Identification and Authentication Failures*
- *A08:2021–Software and Data Integrity Failures*
- *A09:2021–Security Logging and Monitoring Failures*
- *A10:2021–Server-Side Request Forgery (SSRF)*

OWASP went a little further and also introduced a Kubernetes-specific *Top Ten* list as of 2022:

- *K01*: *Insecure Workload Configurations*
- *K02*: *Supply Chain Vulnerabilities*
- *K03*: *Overly Permissive RBAC Configurations*
- *K04*: *Lack of Centralized Policy Enforcement*
- *K05*: *Inadequate Logging and Monitoring*

- *K06: Broken Authentication Mechanisms*

- *K07: Missing Network Segmentation Controls*

- *K08: Secrets Management Failures*

- *K09: Misconfigured Cluster Components*

- *K10: Outdated and Vulnerable Kubernetes Components*

The lists by and large match up, but both apply to a Kubernetes-based IDP. Instead of taking these lists as comprehensive guides to create a security posture, these should be considered bare-minimum items to address in the security posture of the IDP but still represent a comprehensive start to your scoping project.

Implementing threat modeling

After you scope security, step two is **threat modeling**. A threat model is a representation of everything that can impact the security of your application or, in this case, platform. Performing threat modeling is an excellent example of how to arrive at the correct conclusions for the security posture of an organization. You can use these *Top Ten* lists to guide your conversations around threat modeling. According to the authors of the Threat Modeling Manifesto (`threatmodelingmanifesto.org`), a threat model should answer the following four questions:

- What are we working on?

- What can go wrong?

- What are we going to do about it?

- Did we do a good enough job?

For example, you can start with: *We're working on the way users authenticate to the IDP.* Then, follow that with *What can go wrong with overly permissive RBAC configurations?* and work through each of these *Top Ten* items in the Kubernetes list. These questions are deceptively simple, but the surface area quickly expands as question two (*What can go wrong?*) and question one (*What are we working on?*) are a many-to-one ratio, as is the same for question three (*What are we going to do about it?*) compared to question four. In any case, where the answer to *Did we do a good enough job?* isn't a decisive yes, then the loop of questions should be repeated. A successful threat model and response plan is conducted in a collaborative way with all stakeholders of the platform involved.

This collaboration on security is one of the most important strategies to successfully center security without sacrificing usability, the ability to accept contributions, or self-service. It is also through a collaborative threat modeling process that sociotechnical risks to security can be addressed. Security to the left doesn't just mean as far as the developer's computer, but the developer themselves. Ensure they're taking appropriate precautions, are aware of how bad actors try to manipulate situations to steal credentials, and are generally following best practices–such as not leaving a company laptop in a car from where it could be stolen.

Common security standards and frameworks

Navigating the world of security standards starts with trying to demystify a language of acronyms. The goal isn't to become a security expert overnight but to instead know what the level of security you need is and ensure the platform does everything necessary to conform to that security. An easy way to approach this problem is to look at the industry your company serves and what security frameworks are relevant. A hospital or large medical group in the US, for example, would need to follow **Health Insurance Portability and Accountability Act (HIPAA)** compliance. As such, any vendor to a similar organization regardless of locale would need to be able to conform to the same. By understanding the end customers and the needs of the development team, a platform team can determine what level of security and compliance is required beyond standard best practices.

Let's do a quick overview of some security standards. This is by no means an exhaustive list but addresses some of the more common standards:

Standard	Locale	Levels	Description	Notes
PCI DSS	International	1-4	**Payment Card Industry Data Security Standard**. Defines security and compliance. Applicable to any company that processes, transmits, or stores credit card information. Levels are based on transaction volume.	Created by credit card brands, not the government.
DPDPA	India	N/A	**Digital Personal Data Protection Act**. Defines how personal data can be processed.	The government of India passed this in 2023, and while it's comparable to the **General Data Protection Regulation (GDPR)**, it also has notable differences.
FedRAMP	US	Moderate, High	**Federal Risk and Authorization Management Program**. Defines the security and compliance required to provide software and services to the federal government.	US federal government; different from state government.

HIPAA	US	N/A	**Health Insurance Portability and Accountability Act**.	A standard set in the 90s, it's had to evolve with technology.
DGA	**European Union** (EU)	N/A	**Data Governance Act**. Defines the policy of the EU for data use and sharing.	Applies to the context of public sector data and data altruism and what can and cannot be shared. Fills gaps in the GDPR standard.
GDPR	EU	N/A	How personal data can and cannot be used.	This was a historic move and radically changed data handling globally.

Table 7.1: Security and compliance frameworks explained

Although the frameworks differ a little and they seek to accomplish different things, at their core, they are the same. Data captured and stored by an application must be secured while in motion and at rest, and it must only end up in expected places that can be accessed by expected users. This is accomplished partially with RBAC, but RBAC alone will not save you. Security standards such as PCI DSS include a physical inspection of hardware, servers, and access to those physical devices and their housing in order for compliance to be certified. Compliance and security are typically coupled but are actually distinct. A system can be secured but not compliant and vice versa. While we won't go into those differences here, as they're outside of the scope of an IDP by itself, it's important to understand that security is more than just ticking boxes on a list. Those boxes should help inform how far threat modeling work must sprawl through the system and how you develop the role of the platform in any regulated industry.

Asset protection

Sticking to the paradigm of the digital space, the asset we'll discuss is your service's data. Most security and compliance regulations focus on the handling of data. You should take this to mean that data is the most valuable asset the average organization has, and it should be valued as such.

Your data has three states: it is either in motion, in use, or at rest. As such, the security for it must address all states. Since the data lives on the platform, it's the responsibility of the platform to own this part of the security posture. It's a rare product where the data doesn't need to be secured with the utmost care, so it's very difficult to go too far in securing data.

Securing data at rest

Your data will spend the majority of the time at rest. Data design and overall database security require very specific attention and regular review, but the high-level concepts for data at rest fall into the following categories:

- Classification

    - What type of data is it, and what information does it contain?

    - How important is it?

    - Physical isolation of data based on regulatory needs or business weight

    - How the data interacts with the system can inform classification

- Encryption

    - Encryption at every layer, both the physical and the digital

- Salts and hashes

    - Not just for encryption but also for compression

        - Not great if you want data kept human-readable

        - Can expand hugely in memory and could create an accidental **distributed denial of service (DDOS)**

- Restricted access

    - Can be adjusted based on the classification

    - Should have a formal review process, roles, responsibilities

- Redundant backups

    - Three is the magic number for high- and low-availability systems

    - The redundancy of data does not need to be uniform; policy design based on the cost of losing the data

- Data retention policies

    - How long do you really need it for? Are there any governing laws?

        - What is the **chain of custody** on that decision?

        - Same as redundancy; no need for one policy for all data

- Do you need it in the first place?

  - Challenge every piece:

    - Bloat occurs when people assume data is necessary

    - Uncertainty makes people ask for more than they need

    - Showing how data adds value and accountability and makes it easy to justify usage and cost later:

      - Aggressively descope unjustified pieces

      - Commit the history of how and why the system was made

        - Onboarding people becomes easy

        - Answering questions about the system becomes easy, including unfair ones

        - The bus number loses its importance

    - What is the business case to keep the data?

      - Does it make money?

      - Does it save money?

      - Will we learn from it?

      - Is it the right thing to do?

      - What risks are we exposed to by keeping it?

    - What is the operational use case? (e.g.: troubleshooting items, access logs, audit trails)

- Hot and cold storage

  - Hot storage is more readily available

    - Needs access rules

  - Cold storage is not readily available

    - Older data that doesn't need frequent attention but is still important goes into cold storage

    - Harder to access, typically gated by higher levels of permissions

Data sovereignty

Many countries are adopting data sovereignty laws, essentially mandating that data created by people within the boundaries of the country does not leave the physical boundary of the country where it was created. This does not always mean that data cannot be viewed (data in use) outside of the country but that the data storage must remain within the regional boundary. This addresses the data-at-rest compliance posture but not the security posture.

Securing data in motion

When data is in motion, it is being transmitted between microservices, which means that it's exposed to the networking of the platform and/or external endpoints. The data will need to be encrypted during transmission, but ultimately, the data needs to be used so that it will be unpacked by the receiving endpoint at some point:

- Limit what data is stored in memory and for how long–this protects the platform health and the data (be smart about your caching)
- Don't transmit data on ports that are widely opened or privileged
- Only log absolutely necessary information about the data transaction, not the transacted data itself (see log sanitization)
- Protect against injection by sanitizing inputs
- Use networking security and best practices for cryptography to prevent **man-in-the-middle (MITM)** attacks and other kinds of attacks

Data in use

Data in use is exactly what it sounds like: data that the system is viewing or changing. When already stored in use, it's either being retrieved from storage or cached. It might be held in memory or be read directly leveraging various read and caching technologies. Data would also be in use when it is initially entering a system. This includes registering new users or storing a new log line. Oftentimes, data in this state may also be undergoing data transformation such as aggregations or sanitization operations that ensure the data can't be used for an injection attack, or even deletion. Securing data in use is just another application of the same principles we use when data is in motion or at rest.

Securing your network

Kubernetes has a pluggable architecture, and while it ships without a networking stack, some Kubernetes platform options will have their own default. For example, OpenShift Container Platform adopted a default **Container Network Interface (CNI)** known as **Open Virtual Network (OVN)**. Aside from OVN, there are other secure and more observable solutions for Kubernetes networking.

Cilium is an example of this more secure networking. It uses **eBPF**, which is a Linux kernel-level technology that allows kernel capabilities to be leveraged by programs running in a privileged context without needing changes to the kernel or a kernel module to be loaded. Cilium's use of eBPF creates highly secure, observable, and performant networking for Kubernetes environments. Cilium brings those kernel-level capabilities into the Kubernetes network layer. The Cilium project website (`https://cilium.io/`) is filled with hands-on tutorials and labs for gaining familiarity with the technology, and it is a property of the **Cloud Native Computing Foundation (CNCF)**.

In addition to the networking technology, the networking topology will play an important role in network security. Networking tools such as firewalls, VPNs, VLANs, routers, switches, and so on may not live on the Kubernetes cluster but play a very important part in the security of the cluster. Whatever the final network topology looks like and how the cluster interacts with the public internet (or maybe it doesn't!), for proper threat modeling and compliance, you will need to be able to observe your network and document your network.

Isolation between pre-production and production environments

A general best practice is that regardless of security and compliance needs, your system keeps production data and access isolated from other environments to ensure data is protected. Data is the most valuable asset most companies have, so the protection and isolation of data is the best way to secure the asset, guaranteeing adequate security and compliance for all companies. Data protection is at the heart of security and compliance. Production data should never leave a production environment, and access to that data must be rigorously gatekept to ensure no bad actors–internal, external, intentional, or accidental–can gain access to that production data. Let's again reference our platform architecture from *Chapter 2*:

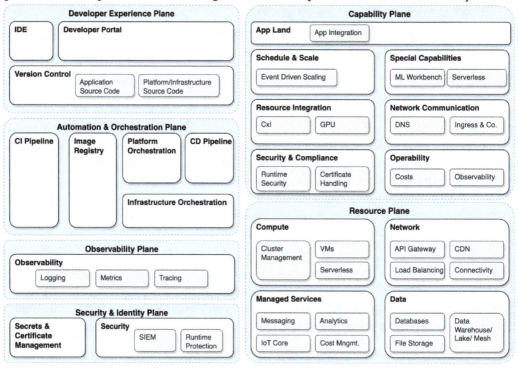

Figure 7.3: Platform reference components

Each white box, even the security-related boxes, must have its own security gates. The ability to manage RBAC for the organization cannot be open for just anyone to change, as an example. It's easy to see how this can quickly spiral into a problem that has spawned an entire field of experts. We won't seek to replace their knowledge and expertise in this book; however, we will share some of the most important aspects of security for an IDP as we see them to get you started on the right path.

An easy win for any organization is to have the staging, development, and production environments all completely isolated from one another. This includes having separate databases with different access rules entirely versus one database with different tables and different access rules.

It is possible to use a single cluster and have network policy-based separations, RBAC that applies to very specific namespaces, and create a similar experience for isolation as you would experience with having multiple clusters, but the cluster's API server, audit logs, etcd, networking, and other cluster-scoped resources will still represent a single point of potential failure of that isolation.

Therefore, for security and compliance, it's best to separate environments completely. By having two distinct clusters, the isolation of the production data is guaranteed and leaves less room for human error to impact that security. From there, you may also have different networking configurations, such as firewall rules that have different allowances, or even environments that are completely disconnected from the public internet.

Secret and token management

In Kubernetes, a Secret is a password, authentication token, environment variable, API key, or similar piece of sensitive data that an application may need to access in order to function correctly or accomplish a task. Secret management becomes one of the most critical challenges to solve when working on a system that relies so heavily on automation as an IDP does. Fortunately, there are patterns and technologies designed to help handle this.

As a Kubernetes-based platform, the built-in Kubernetes functions for secret security are the key starting point. It's important to know that default behaviors are not secure. Secrets are stored similarly to ConfigMaps and are not encrypted. They are encoded, but the encoding is only base64. This is fine for development environments but not so good for production environments. You can, however, encrypt secret data at rest without installing any third-party applications.

For further reading on encryption at rest, and some examples that could be used against a test cluster, please see the *Data encryption* section of the Kubernetes documentation: https://kubernetes.io/docs/tasks/administer-cluster/encrypt-data/.

Secret management is often fairly large in scope, and once you have more than one cluster or more than one environment, it becomes difficult to manage all the secrets required for the IDP to support applications. As such, the use of secret storage software such as HashiCorp's Vault or password managers such as Bitwarden or 1Password has become the industry standard. A standard best practice for landing secrets in a cluster automatically is to store in a code a reference to the secret and then have a logic of some kind that knows how to look up the secret by its reference and fetch it for the application. As such, a common pattern for applications to pull in secret information via references is to leverage a sidecar model.

In a sidecar model, one of the containers in the Pod has the specific job of fetching and making available secrets for use by the main application container upon Pod start. The secrets are then placed into a storage volume and read by the application when it needs them.

A sidecar model would look like this, where the Pod contains both the application and the sidecar:

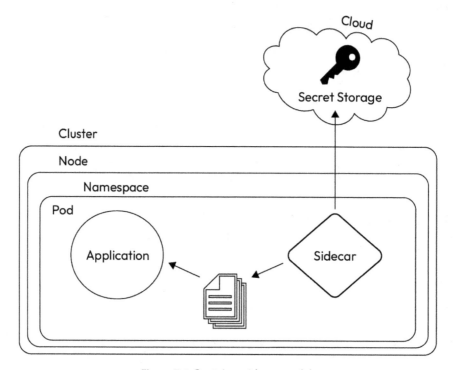

Figure 7.4: Container sidecar model

However, unless the sidecar (or another service) that does the lookup is running on a loop, this is only solved once and doesn't address environments where secrets may be changing on a regular cadence. This is where FOSS solutions are an excellent option. The **External Secrets Operator (ESO)** is a FOSS offering that enables this. The project is a Linux Foundation property. For a full overview of how it works, visit `external-secrets.io`.

Their reference architecture diagram is this, which is pretty much identical to our sidecar reference:

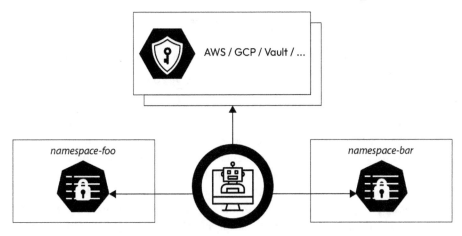

Figure 7.5: Reference architecture for ESO

Essentially, secrets are stored in a password storage that lives outside the cluster, and then the operator is able to access them. The secret references exist as **Custom Resources (CRs)** that the operator knows how to interpret and action the retrieval for. If the secret changes, then the new secret is automatically applied by the operator to the cluster, meaning there is only ever one **source of truth (SOT)** for secrets in the IDP.

Of course, the authentication secret to the storage must be supplied initially for the operator to work. This can pose a little bit of a Catch-22 as whoever sets up this operator will need the ability to provide it with credentials to access the off-cluster secret storage.

ESO is especially useful in highly regulated environments where the contents of secrets, such as certificates and tokens, are rotated frequently. This allows for the rotation to happen at the source but automatically propagates out to the necessary environments.

ESO is not the only open source project for secret management on a cluster; there are several others incubating in CNCF, and all are worth a review. Choosing the correct tool for your organization will need to come after weighing a list of pros and cons, but the pattern leveraged by these tools represents the best practice you should seek to emulate.

Sanitizing logs

Any logs the platform generates should be sanitized. This is true for applications as well; however, that exceeds the boundaries of the platform. Logs are part of the data and, therefore, assets that the security of the platform needs to protect. In addition to the storage and transmission topic we covered previously, data sanitization is a key part of ensuring that even if a bad actor gains access to logging data, they cannot use it to further compromise the system. Code quality checking tools such as SonarQube (`https://docs.sonarsource.com/sonarqube/latest/`) are excellent for detecting any sensitive data going to the wrong location, allowing for remediation of the issue before a security incident can occur.

Logs that are sanitized are cleared of passwords or tokens in their body. For example, when logging requests to an API, how the request was authenticated (for example, bearer token) could be logged, but the actual token itself needs to be scrubbed. Where sensitive data is captured or stored, it should be salted and hashed.

These platform logs should also be kept as clear as possible of **Personally Identifiable Information (PII)**. This type of data is typically not required, so storing it creates an unnecessary risk surface area.

Log sanitization also includes retention policies. Just as application data should be lifecycled and destroyed after its **time to live** (TTL) has expired, so too should log data. Logs become progressively less useful over time for the platform team as the platform continues to run, but the information contained within them can retain value where PII cannot be stripped out due to business needs. As such, keeping data around with value to bad actors presents an unnecessary risk that even the use of cold storage doesn't mitigate. When to destroy data is ultimately a business decision, and data transformation could be done to further anonymize data so that only the PII is destroyed if there's a compelling reason to keep platform metrics around forever.

Secure access

We've covered RBAC a few times already. It's explicitly named as one of the OWASP *Top Ten*, so it's clearly important to the security posture of an organization for access controls to be correct. One method of ensuring this is to create service accounts. These accounts are non-human identity accounts that can be leveraged by workloads on the system. Just as with a human, the authentication of a service account requires a token, and that token should be regularly rotated. By splitting access types into human and non-human, you can leverage the **principle of least privilege** (PoLP) to ensure that a human or a workload has the permissions it needs but none of the permissions it does not.

Least privilege should not just apply to workloads; it should apply to people too. When evaluating what permissions users should have, there are a few things to keep in mind for their access. Next, we'll define some high-level best practices for you to investigate:

- Single-user, multi-RBAC:
    - User has independent RBAC roles for staging and production.
    - Staging should maintain parity with production so that actions performed in one environment will prepare the user to do the same in the other.
    - May have break-glass procedures to gain higher levels of access. This access, if it exists, must be auditable, meaning all the facts about it are logged.
- GitOps for security :
    - Can manage RBAC.
    - Reduces the need to grant direct access to the cluster.
    - Does become a SPOF.
    - Centers security left.

Audit logs

What is an audit log? An audit log is the record of actions the Kubernetes API server sees. This means every change in the Kubernetes cluster, both automated and human-initiated, from logins to Pod scheduling, is tracked in audit logs. If there's identifying information, it's going to be in your audit logs. This is because it is a critical path to know who did what action, where, and when, reviewing audit logs for both incident resolution and security **incident response (IR)**. Audit logs shouldn't have just the API request, but if there was a payload – as might be expected with a method such as PUT or PATCH – that payload should be logged as well. Credentials should not be logged as PII and should be omitted in the vast majority of cases.

Leveraging audit logs to determine anomalous behaviors can be done with some base-level observability implementations and corresponding alerting. When defining the platform, you should have come up with user stories and critical user journeys. In these exercises, you looked at what your users will do and expect to be successful with. But did you look at what your users would not do or should not do and expect to be successful?

Behaviors found in audit logs that defy the norms of the platform users are the simplest way to define alerts around potential anomalies. An example of this would be a very high number of 403 errors or a significant number of requests from an IP address outside of a specific **Classless Inter-Domain Routing (CIDR)** range.

Other items automated detection might look for include the following:

- Detecting unusual or invalid user agents or bots
- Multiple user sign-ins or sessions from different locations

In general, events that violate known norms should probably reach human eyes automatically. However, most of these still require human review as they do not always indicate a security incident. They may indicate a software issue or an event that is necessary but was not factored in when the alerts were designed. Alerts should not become overly noisy. False signals can cause harm, especially when they page your team in the middle of the night. If false signals fire too often, they may soon be ignored by engineers looking to maintain a lower cognitive load.

We've covered the basics of security so far and have a long way to go. Now that we've learned about some general topics, let's get more specific.

Looking at SBOM practices

Open source tools, libraries within programming languages, package managers, and container images are the building blocks of the modern application and also introduce a unique set of challenges when it comes to securing your **software supply chain**. This is what we affectionately call the supply chain security conundrum. How do you maintain a good security posture when you don't own all of the code that needs to be secured?

If we represent the supply chain visually, it'll have some unknown people (we'll call them actors) contributing to an open source dependency and another likely known actor contributing to your code base more directly. This is an extremely simplified drawing (there are probably 10 boxes missing here) but it should help you get the point:

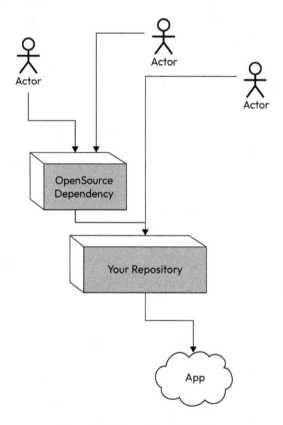

Figure 7.6: Example supply chain

Your software supply chain is everything and everyone is involved in releasing your app. When we look at how to maintain security, we have to break down our application and infrastructure topology. An SBOM is an important tool for tracking and managing the risks of a project.

After the US government issued an executive order in 2021, these documents became mandatory for many companies. The essential premise is that a company knows what the software they're building depends on and where the dependencies of the software originate. An SBOM is generated at the time a piece of software and all of its dependencies are bundled up for release. While not strictly necessary for every company, these represent a best practice as, when paired with a scanning tool, they can assist with audits and understanding the surface area of a risk if a Day 0 vulnerability such as Heartbleed or Log4j is found again.

An SBOM is usually generated at build time in the CI pipeline. **Syft** is a pretty common SBOM generation tool and often is paired with the scanner **Grype** since they're both free open source tools from Anchore. Cisco's Open Source Program Office also recently released an SBOM tool called **KubeClarity** (`https://github.com/openclarity/kubeclarity`), which can use multiple SBOM and scanner tools in concert to provide the most complete picture of the software and its surface area for risks.

SBOM generation tools are still fairly new, and so they're not yet perfect. It's possible for one tool to miss a package the other tool detects and vice versa. For understanding your security posture, less is not more, so gaining the most complete picture possible is the most important part of staying on top of security.

How to use an SBOM

An SBOM is more than just a checkmark on a list of requirements from a customer or the US government. It's also an effective tool for vulnerability detection and response. In *Figure 7.6*, we show how your application likely inherits code, and therefore vulnerabilities from dependencies and libraries it may leverage that are open source. Those dependencies are difficult to track, which is why an SBOM can act as the ledger for your system. This means that down the line, if a significant security vulnerability is announced, you can quickly cross-reference that announcement with your SBOM and understand in a timely manner if your software is vulnerable. This can be done by looking at an already created report and also by simply running the report again. If your report generation is paired with scanning, the scanning tool should pick up on the new vulnerability as soon as it's entered into the critical vulnerability registry.

Getting an SBOM for a GitHub repo

An easy way to see an SBOM for a GitHub repo is to curl the GitHub API for the repo you want to investigate. For the sake of a quick example, we'll walk through how to do this and how to interpret the results.

GitHub SBOMs are in a format known as SPDX; you can learn more about that format here: `https://spdx.github.io/spdx-spec/v2.3/introduction/`.

To get an SBOM, use the following code block in your terminal to curl the GitHub API. You do not need to authenticate for this to work, although you can do so. Replace the `$REPOSITORY` and `$OWNER` variables with your desired repo. For the sake of the example, we'll look at the `tag-security` CNCF repository:

```
curl -L \
-H "Accept: application/vnd.github+json" \
-H "X-GitHub-Api-Version: 2022-11-28" \
https://api.github.com/repos/$OWNER/$REPOSITORY/dependency-graph/sbom
```

The JSON return will be somewhat lengthy, so let's just look at a subset of the SBOM we've received from curling (`https://api.github.com/repos/cncf/tag-security/dependency-graph/sbom`):

```
{
  "sbom": {
    "SPDXID": "SPDXRef-DOCUMENT",
    "spdxVersion": "SPDX-2.3",
    "creationInfo": {
      "created": "2024-08-16T05:37:53Z",
      "creators": [
        "Tool: GitHub.com-Dependency-Graph"
      ]
    },
    "name": "com.github.cncf/tag-security",
    "dataLicense": "CC0-1.0",
    "documentDescribes": [
      "SPDXRef-com.github.cncf-tag-security"
    ],
    "documentNamespace": https://github.com/cncf/tag-security/
dependency_graph/sbom-0a6b74785f7954ee,
```

At the top of the output, you have some basics about the SBOM, including how it was generated, when, and pertinent data licensing. This just tells you about the repo you've analyzed at a high level. The next section is `packages`, which contains all the software dependencies it includes and the relationships:

```
      "packages": [
        {
          "SPDXID": "SPDXRef-npm-babel-helper-validator-
identifier-7.22.20",
          "name": "npm:@babel/helper-validator-identifier",
          "versionInfo": "7.22.20",
          "downloadLocation": "NOASSERTION",
          "filesAnalyzed": false,
          "licenseConcluded": "MIT",
          "supplier": "NOASSERTION",
          "externalRefs": [
            {
              "referenceCategory": "PACKAGE-MANAGER",
              "referenceLocator": "pkg:npm/%40babel/helper-validator-
identifier@7.22.20",
              "referenceType": "purl"
            }
          ],
```

```
        "copyrightText": "Copyright (c) 2014-present Sebastian
McKenzie and other contributors"
        },
```

Looking at just one of the packages in the SBOM, you have the package information, the version in use, and licensing and copyright information. The MIT license means that the package is open source, but the copyright information indicates who has been maintaining the package and, essentially, prevents the name of that package from being used by another software project. IBM explains the reason for copyright here: `https://www.ibm.com/topics/open-source`. Also included in the output are the label supplier and download location. For these two fields in this example, you'll find the metadata says `NOASSERTION`. As explained in the SPDX documentation (`https://spdx.github.io/spdx-spec/v2.3/package-information/`), this should be used in the following circumstances:

- The SPDX document creator has attempted to but cannot reach a reasonable objective determination

- The SPDX document creator has made no attempt to determine this field

- The SPDX document creator has intentionally provided no information (no meaning should be implied by doing so)

For other packages, these fields may have additional data about the package or may also have no assertions. After the list of packages, in the same order are the relationships of the packages to the repository:

```
    "relationships": [
      {
        "relationshipType": "DEPENDS_ON",
        "spdxElementId": "SPDXRef-com.github.cncf-tag-security",
        "relatedSpdxElement": "SPDXRef-npm-babel-helper-validator-
identifier-7.22.20"
```

This is pretty straightforward in this example as it's saying the primary element, the tag security repo, depends on the `npm:@babel/helper-validator-identifier` package at version `7.22.20`. Additional metadata could be supplied by the SBOM creator but hasn't been in this case. More information on these relationships can be found here: `https://spdx.github.io/spdx-spec/v2.3/relationships-between-SPDX-elements/#111-relationship-field`.

Again, an SBOM is a pretty simple tool; it creates a ledger that tallies what an application or codebase is composed of. By itself, it doesn't do much, but as part of a toolchain, it goes a long way to understanding your systems and the surface area of risk for vulnerabilities their dependencies present.

Keeping on top of vulnerabilities

Software vulnerabilities are typically called **Common Vulnerabilities and Exposures**, or **CVEs**. The US **Department of Homeland Security (DHS)** maintains a public CVE registry (`https://www.cve.org`) where you can keep abreast of what exposures are known, and this can be referenced against your software and applications to check for known CVEs and your exposure. While a platform team can build a bespoke service to determine if there are CVEs in the systems, there's no need to do this as plenty of FOSS tools exist to do this for you.

You can check for CVEs in GitHub via the following:

- **Dependabot**, a GitHub bot that scans your repo and raises pull requests to proactively address CVE by bumping packages to known safe versions
- **Snyk**, which validates pull requests to ensure they do not introduce any new vulnerabilities
- **CodeQL**, similar to Snyk, evaluates the content of pull requests to ensure they do not introduce vulnerabilities

For keeping track of vulnerabilities during runtime, regular scanning of an image registry such as Harbor or a tool such as Trivy targeting your most critical environments (minimally) will help you to keep on top of vulnerabilities in your environment.

Since SBOM generation and vulnerability detection are usually part of a CI pipeline, let's move on to discuss the rest of the pipeline and how you can ensure security in your CI/CD process.

Understanding pipeline security – what you have to consider to secure your CI/CD pipelines

Assuming the platform team does have influence or jurisdiction over GitHub or other source control repositories leveraged by the company, then the security of the CI/CD pipelines end to end becomes a major part of the IDP security posture.

Securing your repo

The security of the code repository is an excellent example of security to the left. By enforcing security norms early and baking them into the way of working on a project, an organization can prevent issues from arising down the line. A secured repository leverages several best practices:

- **Write protect main branches**
  - You can use private Git repositories and self-hosted Git if you need extra security
- **Require signed commits**
  - This validates the identity of the commit author
- **Pre-commit webhooks**
  - Used to validate that no secrets are accidentally committed

- **Mandatory peer review**
  - Including signoff by code owners
- **Automated validation of pull requests**
  - Dependency security scans
  - Tests – should include validation of security and access
- **Continuous scanning**
  - Scans for accidental commits of passwords, tokens, or other secret data

This list is by no means exhaustive but should start any organization off on the best path for ensuring their source code is secured against bad actors and human error.

Securing GitOps

GitOps was covered in depth in *Chapter 5*, so we'll simply review it very briefly due to its importance for platform security. GitOps is loosely defined as an automated process that validates the SOT (Git or other version control), which defines the desired state and the actual state match. It does this through one-time deployments or changes but also through an automated reconcile loop that actively checks for changes in the desired state and acts upon the actual state to make it match, or detects a change in the actual state and makes changes on the system to put it back to the desired state. The current largest open source GitOps project is Argo CD, a CD tool that is a property of CNCF. Within Argo CD and other GitOps paradigms, there are a few best practices to note for guaranteeing a more secure environment.

For GitOps, there are two models of propagating changes; one is a push model and the other is a pull model. In a push model, the GitOps system receives changes as pushes to its API endpoint. It goes without saying that those pushes should be properly authenticated, but we'll say it anyway to ensure there's no confusion. After receiving the push, the GitOps system processes the changes and then takes action. That action is probably promoting new software packages to production, but it could also be configuration changes. However, the source of the push is not typically validated with this model, and Argo CD doesn't have a configuration to validate it. This creates an attack vector because if the credentials are compromised, a bad actor can push changes through the CD system, which can open up additional entry points.

A pull method is just the opposite. The GitOps system reaches out to the SOT using its authentication and then reads the changes in from the endpoint. Since the source is a known safe source, the application of those changes via automation is considered a safe action. This is not to say a pull model isn't without risks. Both push and pull models can be subjected to MITM attacks, but in a properly secured network with the correct cryptographic implementations, these risks should be significantly limited.

When your GitOps system is taking action, it should do so in the most secure way. We've already discussed the uses and value of a service account, so it should come as no surprise that your GitOps systems should be expected to leverage service accounts. Those service accounts should also have the minimum amount of access possible while still being able to accomplish the goals of the GitOps implementation in the platform.

Security and GitOps may be an important factor in selecting the GitOps working model for your IDP, and we hope these guidelines are useful when building out GitOps for your organization.

Now that we've covered securing the delivery of the application, let's build off our secure foundation and take a look at application security.

Understanding application security – setting and enforcing policies

Security is a moving target, and as technology advances, attack vectors increase and attackers become more and more sophisticated. Due to this, the process and ceremonies the team maintains around security are even more important than technology itself. This is not because technology doesn't matter but because good discipline and the habit of a strong process allow technologies to plug in and out as the industry evolves.

One aspect of good discipline for security is maintaining accurate documentation and architectural diagrams. If a significant change is made to the application architecture, then that can change the risk surface areas and attack vectors. For example, an undocumented or underdocumented dependency on a library or a network port could result in exposure to a vulnerability that may be more difficult to determine.

Foundational application security

In *Chapter 5*, we discussed building and delivering images and artifacts. The described method of semantic versioning for an application represents a best practice for creating software but not for using it. When a release is built using Git for source control for an application, the release, in addition to a human-defined version, actually gets a SHA-256 signature thanks to the modern functionality of Git. Unlike the version numbers, which can be reused, the **Secure Hash Algorithm** (**SHA**) is a signature for that exact build of the software and it is always unique. As such, it's important for security best practices to use the full SHA address for the images used by the platform as opposed to the image version.

Here's an example of the docker pull command using the image version and SHA:

```
docker pull quay.io/keycloak/keycloak@sha256:520021b1917c54f899540af-
cb0126a2c90f12b828f25c91969688610f1bdf949
```

In addition to image versions, there is another identification method known as **floating tags**. A floating tag is a pointer and applies an arbitrary value designed to be reused to a release image. The most common tags are `latest`, but they can be pretty much anything since there are only industry norms and no technical limitations. Since a tag can be later re-pointed to any release image, it's not a secure way to pull in a dependency. Risks of using floating tags include the following:

- Unintentional and untested/vetted updates to application components.

- Limits the ability to immediately know what's running where and when.

- Unvetted external software packages can automatically propagate, creating exploitable backdoors for bad actors. This is common if a Pod restarts and `latest` is used in the Dockerfile for the Pod.

There are a few reasons to use a floating tag or to use a semantic version release tag instead of a precise SHA. It's easier to test the continued supportability of a dependency if you're testing regularly against the latest build. The flexibility that comes from always using the latest version of a known good package may be desirable in a development environment and then could be pinned to a precise version tag or SHA when moving to production. Validating against floating tags can allow a team to be more responsive to security vulnerabilities as well, so a risk assessment of each dependency and a policy should be set to determine what the norms of the application will be.

A typical way of setting a policy would be to use tools such as open source policy agents to catch any deviation from the defined norms and prevent those deviations from entering restricted environments. **Open Policy Agent** (**OPA**) and **Kyverno** are both open source options that are free to use. For both tools, you leverage **Policy as Code** (**PaC**), which is easily reviewed and kept in source control.

OPA and Kyverno are two examples of OSS that might be leveraged for security, but they're not the only tools in the open source ecosystem.

FOSS for platform security and how to use it

FOSS projects, both within and outside of the Linux Foundation or CNCF, exist in great abundance to help you manage the security posture of your platform. Previously mentioned projects such as Harbor and Trivy are two among a large number.

When comparing your security needs, such as ensuring the OWASP *Top Ten* is addressed for available open source projects, you'll find a tool to help you address each item on the list.

Patterns and tools for managing security

The platform can only do so much to manage the security posture of the company. As such, it needs to provide a strong foundation for security by providing useful integrations, taking a security-first approach, and, as we discussed in previous chapters, being open to contributions from the developer community it supports. When you've determined what levels of compliance are necessary, then you can start looking at what mistakes might lead to a failure in your compliance and security posture by conducting process-based security reviews and technology-based security reviews:

- A **process-based review** would mean that a person is reviewing and assessing compliance and security. This could be an audit done to seek certification, or it could be an internal review conducted regularly to ensure best practices are still in effect and guidelines are up to date.

- A **technology-based review** would leverage software to automatically and likely continuously review and validate security and compliance. SBOMs and CVE scanning are examples of technology-based reviews of software builds, and policy engines such as OPA or Kyverno can help with the automated governance of the IDP. Technological solutions can also go the additional step of helping to detect anomalies and events that may represent a security incident. The CNCF project **Falco** (`https://falco.org`) does this. It has several key features, but importantly, it detects if access levels are escalated, which may represent a bad security and compliance posture or could represent a bad actor who has gained access to the system.

For any company looking to demonstrate a level of security and compliance in accordance with a governing framework, that compliance framework will help to define the frequency of the process execution and pair that guidance with regular audits to help ensure the platform isn't slipping.

What would our fictitious company do?

Our fictitious company Financial One ACME is a long-standing financial institution that has been working on its cloud-native transformation in order to remain competitive with younger Fintech companies. As a financial institution, they have an inherent goal to minimize risk. They are also subject to regulatory restrictions, including PCI DSS.

Additionally, as they are a bank and must protect both monetary assets and customer data, they have modeled potential threats to their systems. Out of a long list of physical and non-physical risks, many security action items came to the platform team to be addressed during the implementation of the IDP. These items have likely been addressed already, but since this IDP is a greenfield application (or brand new), it needs to address the compliance requirements afresh.

One item that came to the platform team for implementation was guaranteeing that all new versions of their software bundle were properly secured.

Properly secured means the following:

- Signed and validated
- Stored securely
- Retrieved securely
- Strict change control
- Audit trail

The platform team took the steps previously outlined to secure their repository and their GitOps system, but those weren't enough to satisfy this requirement. For this reason, the platform team decided to leverage an internal image registry where all software commits and packages are signed. The CI/CD pipelines create images based on those artifacts and place them in the private registry, where platform applications consume them.

The image registry selection would look like one of a few options:

- Pay for an image registry vendor that helps them maintain their security posture via scanning and provides them with trusted **Universal Base Images** (**UBIs**)

- Host their own image registry such as Harbor, a CNCF-graduated project with an image registry that *signs*, *stores*, and *scans* container images (`https://goharbor.io/`) that they would populate with approved images

- If they are using OpenShift, there is already an image registry as part of the cluster topology, and the platform team would supplement it with scanning tools

They've further added an admission webhook to ensure that no container is based on an unapproved image. This use case is very straightforward: a validating webhook checks to see if the container image matches the defined expectations. If not, the workload is rejected and the Pod will not start. While the team could build their own admission webhook service, they would likely choose OPA. While it is not the only open source project to provide this functionality, it is the only graduated project, making it the safest choice to use in production.

Sysdig made an open source version of such a webhook that also mutates or modifies Pod configurations to have their images use the full image SHA versus a release tag, which can be found here: `https://github.com/sysdiglabs/opa-image-scanner`. Both of these webhooks represent absolute best practices for security and compliance and are a smart move for an easy security win for any platform engineering team.

However, as important as these restrictions are, they can negatively impact innovation by removing flexibility from the IDP capabilities. For this reason, the platform team has decided to limit these restrictions to the production environment. Doing so allows the development teams to experiment in their dev environments without risking anything that shouldn't accidentally make it into production environments.

One strategy to manage dependencies where you don't want to pull source code from untrusted locations is to use vendoring. Vendoring is the process of actually copying the code from the open source library you'd otherwise import into your code base. During vendoring, changes may be made to that code to make it more secure, such as enabling **Federal Information Processing Standards** (**FIPS**) mode. FIPS compliance specifies the strength of the encryption of data over **Secure Socket Layer** (**SSL**) and is generally a best practice.

Another action item that might come to the platform team would be to be the first point of contact in the event of a suspected security breach. This means having a way to immediately react to reports of an issue. Since the platform is responsible for so much of the company infrastructure, from development to production capabilities, the team will need to be able to quickly engage in order to mitigate damages caused by a bad actor.

Such **IR plans** (**IRPs**) should be crafted and regularly tested. Similar to a **disaster recovery plan** (**DRP**), regular testing of the plan and the ability to respond to a real security breach can mitigate damages.

Early detection is another key capability required by the platform. Analysis of audit logging is key to understanding who got in (or whose credentials were compromised) and what was done once the bad actor gained access to the system. Audit logs can also be used for proactive detection as they can be leveraged by **machine learning** (**ML**) models for anomaly detection, which can highlight a security breach faster than a human can find it. Additionally, using a highly observable cloud-native networking solution such as Cilium can help to identify and track bad actors. While some hardcoded observability implementations can attain the same outcomes, it does need to be manually kept and maintained, whereas ML models may have some more inherent flexibility due to their self-learning nature.

Neither method is perfect, so when determining how to implement automation around threat detection, an organization will have to make a judgment call on benefits, trade-offs, and team capabilities. Choosing the path forward here for our fictitious company would be a build-versus-buy conversation that could very well result in a buy decision due to the size and complexity of the task and the highly secured environment they're meant to maintain.

Speaking of that highly secured environment, in order to demonstrate to an auditor the PCI compliance of the organization with the new IDP, our platform team will need to be able to produce a drawing of the network architecture and explain the contents of that drawing to the auditor.

Any compliance auditor will want to verify that the dev and production environments are sufficiently isolated, whether that's physically true or via networking implementations.

Finally, while the platform team at Financial One ACME will be making sure they follow all known best practices for build time, they'll implement tools to ensure their runtime is equally as secured. It's very likely that the only users with permissions to create Pods in the production environment would be those service account users, ensuring that Git remains the SOT, and GitOps can be leveraged to codify the platform's security posture.

These example responses cover only a subset of the security and compliance that would be required for a financial institution such as a bank; they're also likely to be subjected to additional government regulations, which would require even further security and compliance risk mitigations. Just as your team will need to, our fictitious company would need to address every line item within the compliance frameworks they're subject to and, most importantly, create collaborative ceremonies to maintain the best security they can.

Summary

To conclude, security and compliance is a vast space with many experts who have published dedicated works on the subject. This chapter should not be taken as being all-encompassing but should have you started down the correct path to define and execute a cybersecurity strategy for your IDP. It's important to know how to keep track of vulnerabilities, and have ceremonies and tooling set up within your organization to catch and surface vulnerabilities in the IDP and the applications it hosts.

While security and flexibility are not natural partners, smart implementations that focus on critical security needs without impeding innovation are the key to providing the developers with the tools they need to be successful and the protections they need to be secure.

Remember – the cost of a security incident can be astronomically expensive and could even result in bankruptcy or trials. While log storage and other security requirements can cost, those costs can be managed and will never cost more than failing to secure your systems. For more on how to manage the costs of your platform, let's continue to *Chapter 8*.

Join the CloudPro Newsletter with 44000+ Subscribers

Want to know what's happening in cloud computing, DevOps, IT administration, networking, and more? Scan the QR code to subscribe to **CloudPro**, our weekly newsletter for 44,000+ tech professionals who want to stay informed and ahead of the curve.

https://packt.link/cloudpro

8

Cost Management and Best Practices

Cloud migration has been the top priority for many IT organizations in the past few years and is still a strategic, relevant direction for the coming years. Running and owning one's own data center is also still a valuable approach; the design and costs for one's own data center capabilities are usually more thought through than with the cloud. In this chapter, we are going to dig deeper into this and will reflect on the necessity of cloud cost management and optimization.

You will learn more about tagging strategies and why they are a viable resource for gaining visibility and transparency in your cloud spending. We will also examine how to define them, best practices, and practical approaches to ensure they are well set. With this as a foundation, we will move on to the four pillars of cost optimization: *processes*, *pricing*, *usage*, and *design*.

To close this chapter, we share practical tips on how to optimize your platform and reduce your costs in the long run, as well as enabling you to provide cost-saving value to your users.

Overall, we will focus on effective cost management and how, as a platform engineer, you realize that. Here's what you can expect to learn:

- Understanding the cost landscape—is the cloud the way to go?
- Implementing a tagging strategy to uncover hidden costs
- Looking at cost optimization strategies
- Autoscaling, cold storage, and other tricks for cost optimization

Understanding the cost landscape – is the cloud the way to go?

Cost management in platform engineering begins with a good understanding of cost drivers within your infrastructure and how some platform components might influence those. Cost drivers are also elements that directly impact the total cost of your operations. Understanding and later identifying these can help in making informed decisions to provide cost-oriented and optimized platforms.

To cloud or not to cloud – that's the question

For the past few years, there has been almost no way around a cloud adoption strategy. During those years, the movement has faced counterweights from companies that have successfully de-migrated, claiming that their on-premise setup is cheaper, better, and all they need. *Repatriation* is what this is called. The numbers of those doing this are not very clear and strongly depend on what is considered part of it. It became very famous as *HEY/Basecamp/37signals*, and their CTO David Heinemeier Hansson stated that *they will save over $7m over the next 5 years* if they do not run on the cloud. A dead stupid side-by-side comparison of their purchased servers that cost them half a million dollars versus their insane yearly cloud spend of $1.9 million makes the math complete *[1]*.

Sure—Basecamp or HEY is not comparable with an enterprise, is it? From my own experience, I can tell that running even larger companies' hardware in their own data center is cheaper than going to the cloud. A key factor is the amount of data and how dynamic and scalable the infrastructure has to be for it. Calculating data in petabytes will quickly turn the cloud into a money sink without an end. A relatively static workload will also drastically reduce the benefits that are included, and it has many other drawbacks. On the other hand, the cloud market is growing continuously every year, with no end in sight. So, to come to a decision on what is the right way to go, you have to consider many moving parts, from a simple cost comparison to available skills and the actual demand of your business. Doing a full assessment of whether the cloud makes sense for you can be a project in itself. But we wanted to give you some ideas on criteria to consider if you are at this point:

- **Cost**: Evaluate initial CapEx or one-time costs for on-premise versus OpEx or continuous, subscription-like costs for cloud services, considering long-term financial impact but also that you will save the most money on cloud computing if you pay for 1 or 3 years upfront. This is almost like buying the hardware.

- **Scalability**: Assess your demand to scale resources dynamically in response to fast changes. A hint here: if the spikes are too fast, a static server might be better for the user experience than waiting for minutes until new instances start.

- **Reliability**: Evaluate uptime guarantees, redundancy, and failover capabilities to maintain continuous operations. Have a look at the history of outages. Some cloud providers frequently have issues.

- **Compliance**: Ensure adherence to legal and regulatory requirements, focusing on data residency and sovereignty. You have to sharpen your point of view on sovereignty before you can decide. Unfortunately, this term has more of a spectrum than a clear list of requirements.

- **Integration**: Ensure compatibility with existing systems and software, and availability of APIs and integration tools. The provider should support **infrastructure as code (IaC)**, especially to programmatically create accounts, users, and permissions.

- **Support**: Assess the availability and quality of technical support, along with **service-level agreements (SLAs)**. Conduct research on other users' satisfaction with the given support; just because you have a contract that someone keeps your system up and helps you doesn't mean that the service is also good.

- **Development of new services**: Evaluate the speed and number of new features the provider gives you. Consider that there can also be too many updates per month but also that there are barely any new features.

- **Geographic considerations**: Consider the proximity of data centers to end users and the regional availability of services. If your company develops globally available **software as a service (SaaS)**, it might be easier to build on a public cloud than to integrate several regional providers.

- **Skill and expertise**: Evaluate the availability of skilled personnel and training requirements for managing the infrastructure. Consider that you still require practically anyone from any IT subject who can do that. What does a short-term plan look like? What is needed in the long run? Forget about cloud computing being easy; most enterprises struggle because they don't get their teams educated.

- **Environmental impact and sustainability**: Consider the provider's energy consumption, carbon footprint, and sustainability practices. How do they manage their wastewater? If they buy carbon offsets, keep in mind that you have to pay for them, as they theoretically increase electricity costs.

Cloud has often one very strong point: almost nothing hinders you from just starting right away. You can find tons of templates, blueprints, and examples. And within a short time, you are up and running, having your first environment available as a platform.

When we opt for the cloud – we have to consider its hidden costs

As we focus primarily on the cloud in this book, we will assume for now that those who go the bare-metal way have by nature a higher cost awareness and sensitivity.

Cloud providers give us services on hand with a lot of *batteries included*. They often come with backup solutions, scalability, **high availability (HA)**, and a centrally managed service. When planning the cloud infrastructure, most of the time, we will find hard facts and best guesses. Hard facts are information such as how many CPUs or servers are needed, which database should be used, whether it is a single node or HA, and so on. However, you should be aware that almost every option you take with a public cloud provider will have some cost impacts.

Big cost drivers that commonly appear include the following:

- **Load balancer**: They are required in almost every architecture, and the same goes for you. As explained earlier, we have possibilities to extend the app services from the Kubernetes namespace to the cloud and even control the network – a major cost trap.

- **API gateway**: The even more evil twin of the load balancer, applied as best practice, becomes very costly, especially for chatty systems and routing heavy communication.

- **Data**: In rest or in transit, data quickly becomes the biggest cost block in a cloud bill. While everything can be scaled up and down, your data has to live somewhere.

- **Backups and snapshots**: They are very much needed for a mature platform; you are required to have a very good backup strategy that is lean but reliable.

- **Scalable managed services**: If you're using serverless, message streaming, or pre-trained AI models—anything that can scale infinitely—it's crucial to set limits to avoid incurring unexpected costs.

Ideal cloud projects—and we have to emphasize here that we are talking about things with a deadline—are strictly opinionated, following the best practices of the cloud provider. Almost every cloud project is not ideal, causing friction in the migration by custom implementations that are needed to somehow make the old world and the cloud provider able to work with each other. These hidden costs accrue through missing skills, because where should the required skills suddenly come from? Getting help from external skill providers would mean they have to understand you first before they can actually help you. However, most cloud projects would be called dead way earlier without external support.

In short, we can say that most services can turn against you cost-wise. As platform engineers, we have to be able to control these elements and create guardrails and limitations. To enable this, we need transparency about costs.

Where to find transparency

Meanwhile, we have a zoo of possibilities for retrieving information about where we spend money and whether this is considered a waste based on metrics such as utilization. All cloud providers offer the possibility of exploring (and exploding) costs, but due to the focus on the infrastructure, details are missing, especially on the application layer. Looking at the following screenshot from AWS Cost Explorer, we get a consumption graph of different service types. We now would need to apply filters to gain more insights and analyze the roots of the costs:

Figure 8.1: AWS Cost Explorer

Commercial solutions such as Apptio Cloudability can gather costs from many different accounts, cloud providers, and environments in a single tool. They apply FinOps logic on top of the given data and come with predefined dashboards such as the unit costs and savings dashboard from the following screenshot:

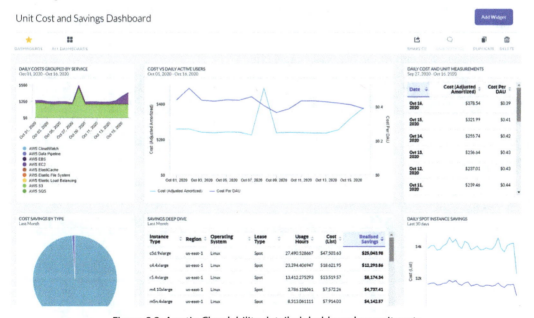

Figure 8.2: Apptio Cloudability detailed dashboard on unit costs

Microsoft defines unit costs as follows: "*Measuring unit costs refers to the process of calculating the cost of a single unit of a business that can show the business value of the cloud.*" What a unit is exactly is up to you. It can be transactions in financial systems, users for a media platform, or anything else you can map different parts of the infrastructure to.

Now, in our platform and open source world, we can also find solutions to gain more transparency, such as Kubecost (with commercial plans) and OpenCost. Both can allocate costs within a cluster and in combination with cloud provider resources. Unfortunately, both lack good analytics capabilities. That's why we often see self-implemented solutions in combination with **business intelligence** (**BI**) tools, or Prometheus and Grafana implementations enriched with some cost data. How good those solutions are depends on the time and money invested.

FinOps and cost management

In recent years, FinOps has gained some popularity and is trying to differentiate itself from classic cost management. Cost management is often outlined as a short-term, siloed, single-purpose activity that should yield fast cost savings. When you, as a platform team, act cost-aware and actively manage your costs, this will also be a sustainable and long-lasting solution. But where FinOps succeeds is in its holistic approach, which includes procurement departments, **business units** (**BUs**), and finance departments, clearly targeting implementing an organizational understanding of cloud costs and their dynamics.

The following overview of the FinOps Framework shows that it follows a similar approach to what a platform engineering team is up to. We can see principles that need to be aligned with your principles; capabilities to be integrated and enabled, which will require your input; and personas—roles such as the platform engineer with whom they collaborate *[2]*:

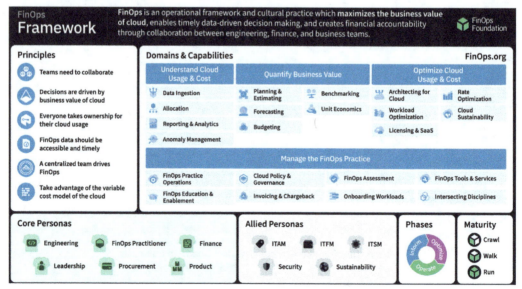

Figure 8.3: FinOps Framework by the FinOps Foundation

As a platform team, you should collaborate with FinOps teams and actively manage their influence on your platform. Potential cost savings must be balanced against your principles, user satisfaction, and developer experience. Data-based recommendations require an architectural qualification and considering alternative options and proposals to realize cost savings.

In the following section, we will discuss how to implement a tagging strategy. Tags are a simple solution for providing additional information to cloud and Kubernetes resources and creating some sort of transparency. Also, many cost management and FinOps tools require tags to deliver better insights.

Implementing a tagging strategy to uncover hidden costs

Tags and labels will most likely be nothing new to you as they are used in a wide range of tools, public cloud, and Kubernetes. We will use the word *tag* also to represent labels, but when using the word *labels*, we actually exclusively write about Kubernetes labels.

Applying and using tags can become an art in itself. If there are too many tags, it can become unclear what information they should attach to the service. Also, having no tags is obviously not helpful. In organizations with many different BUs and departments involved and complex release mechanisms, tags can become overloaded or just a collection of abbreviations. The point is that we need tags to gain transparency on services to whom they belong, maybe indicating different service levels or security classes and becoming in the end an anchor to match those services to cost structures and resource utilizations. So, tags allow a more precise analysis of cost-related resources and the cost cause.

Using tags for a purpose

Tags can target different target groups and users. Depending on which subdomain you are asking about, they should include organizational, operational, security, or even architectural information. We can also think about potentially helpful tags for us as platform engineers.

Tags for organizational information can define which department, role, or team is responsible for the service, whether it is an internal- or external-facing solution, and whether there are any country, compliance, or governance restrictions.

Looking into the operational space, common attributes are service level, scheduling times, and maintenance windows but also very technical information that might be introduced by third-party tools; for example, via Cloud Custodian that implements certain cloud policies *[3]*.

For domain-specific areas such as security or a platform team, we can think of tags and labels as defining a required isolation or data protection, or else like a typical scaling pattern, top and bottom spikes, or identification for a microservice architecture and its belonging to certain components and a structure.

Our focus will be on relevant cost tags. Those are often similar to organizational tags or related to the previously mentioned operational or domain-specific tags. Clarifying organizational belonging or providing insight into why something scales so drastically are key enablers for cost management. These tags help provide a clear view for projects and teams that have to raise cost awareness.

Some sources also point out the use of tags for access management. This should be handled with care for several reasons. If you have **attribute-based access control (ABAC)** and rights management, tags can be a viable and simple approach to achieve that. This comes with the downside that you turn a pure information source into a security-critical element that requires it to be protected and double-checked when used. Mixing this with its information character might complicate the usage of tags and most likely will be a blocker for proper usage. We said before that tags can be used to carry security information. However, there is a difference between providing information such as risk classifications or security levels and an access-granting capability.

Therefore, it is important to, as with our platform, define the purpose of why we want to do tagging. To randomly use and set tags as much as you can imagine can be better than nothing but can become problematic in the future when it is unclear if the tag is now critical or not.

Tag and label limitations

Unfortunately, we have two widely appearing issues:

- Tags don't have a standard, which means there are many variations of their length, number of tags, allowed characters, or case sensitivity

- Organizations tend to have multiple environments across many different providers and platforms

So, to define a tagging strategy in the next steps, you have to know the smallest number you can reliably build on. A tagging strategy will introduce a standard, which has to be applicable on any platform. Defining multiple different approaches per platform will lead to confusion and errors.

As an example, we will compare different providers and platforms:

	AWS	GCP	Kubernetes	Azure
Maximum number of tags per service	50	64	No limit specified	50
Max. characters tag name length	128	63	63	512
Max. characters tag value length	256	63	63	256
Special characters	alphanumeric characters, blanks and + - = . _ : / @	alphanumeric characters, - and _	alphanumeric characters, - . _	alphanumeric characters, <, >, %, &, \, ?, / not allowed
Case sensitive	Yes	Yes	Yes	Yes

Table 8.1: Tag limitations of cloud providers and Kubernetes

While 63 characters as a common direction looks good, we have seen in the field that for many organizations, even 256 characters weren't enough. This is just another example of why a tagging strategy must be thought through.

Also, some structures don't work across the different providers. Assuming our Financial One ACME company utilizes AWS and GCP, including its managed Kubernetes service for the platform, why does Financial One ACME go to the limit of the maximum allowed number of tags? The AWS team discovered this issue first and found out it can use longer tags and some special characters that allow a sort of nested structure. So, they started to combine different tags into one with the following outcome:

```
department-responsible=financial-one-ACME/domain-sales+marketing/
stream-customer-management/squad-frontend/operations-squad-lazy-turtle
```

The value has 112 characters and contains / to represent the hierarchical steps and + to represent *and*. The team is happy with the result and pushes this new tag to the shared repository. Sometime later, the platform team receives complaints that in the sales and marketing IT department, there is a problem with the deployments, and the newest releases aren't getting rolled out due to some labels.

Besides that, you will find other limitations such as the maximum number of tags per account or subscription.

Defining a tagging strategy

As explained earlier, tags can have many different purposes; primarily, we have to balance between technical tags and business tags. Business tags can be categorized into organizational and cost tags. In the end, many tags tend to have overlapping information. Therefore, to find the right approach, it often makes sense to set some basic rules and boundaries for the definition. In the second step, you have to coordinate between operations, platform, and development to find the right set of needed tags from the technical side before moving on to set organizational tags. This direction makes sense because, often, required tags for operation also contain information about the organization.

Let's start with some common ground rules:

- Tag names and values *must* be shorter than 63 characters.
- Tags can be only alphanumeric.
- Tags can only have - and _.
- Focus on tag value and its content as you usually filter and search for this.
- Tag names are more relevant for sorting and grouping.
- Don't write **personally identifiable information** (**PII**) into a tag.
- Establish one case style. Businesses might prefer the Pascal style, while in tech, the kebab style is more common.
- It is better to use more tags than less, but try to keep 10%-20% of available tags free.

Based on those rules, we can create the next relevant elements. First, we need to think about tagging categories for these to become more fine-grained than those we have discussed so far. A good practice is to think of the following categories:

- Ownership—departments, teams, organizations, streams.
- Environment—whatever your staging environments are.
- Project or product—project or product cluster to identify which components belong together.
- Cost center—who is accountable for it? Are there any shared costs that come on top?
- Compliance—regulatory restrictions and requirements, policies, and compliance rules.
- Operational—life cycles, backups, maintenance windows.

From here, we have to work on a naming convention. As mentioned earlier, it's important not to overload a single tag. Respecting the given rules and limitations and including the nature of many software and system components, a tag convention could look like this:

- `department=public-relations-content-management`
- `owner=department-public-relations`
- `data-classification=personal-identifiable-information`

In some languages, it is preferred to follow a Pascal (`Hello-My-Name-Is-Pascal`) or camel style (`i-Am-A-Camel`). The style of writing everything in small letters is called the kebab style. You will have to define those patterns in a very clear way. How many characters should the name have and how many words? Is it more descriptive or just matching what's been identified? The same counts for the value. Ensure that those conventions are documented, communicated, and included in the onboarding or training for your engineers.

Now, the last step is to start building up a *tag catalog with these ground rules*. It is recommended to create those catalogs so that you have a space to describe the meaning, purpose, and addressed group of the tag. This also helps you to keep tags short, as you will not need to add more description to the tags' value. The following table is a simple example of a tag catalog. It can be extended by the unique demands of the organization. It is important to offload into the catalog any information that you might also need to include in the tag but that makes it artificially long and unreadable:

#	Tag Name	Expected Value	Meaning	Purpose	Tag Stakeholder
1	depart-ment	Department handle or code; for example, DE-22-P	BU or the area the app belongs to	Clarify the owner of the application	Application owner; operations

2	applica-tion	Application name; for example, `inter-nal-cms`	Name of the application	Identify the application	Application owner; operations; architect
3	mainte-nance-al-lowed	Day and time when maintenance can be executed; for example, `sunday-0800-1230`	The tag identifies the day or days as well as a time-frame written as `0800`, which means 08:00 a.m.	Defines when the application can be turned down to update/patch or release a new version	Operations; first-level support

Table 8.2: Tag catalog example

Tagging automation

It would be impossible to tag all resources manually, especially on an IDP with many moving components. Certain tools can help us tag resources automatically when they are created or even "patch" them afterward. As we have seen before, on the platform, we have to differentiate between the infrastructure the platform runs on and the user perspective of the workload they run on top or manage from the cluster.

When you manage your infrastructure with tools such as Terraform/OpenTofu, you can tag components within your code as follows:

```
resource "aws_ec2_tag" "example" {
resource_id = aws_vpn_connection.example.transit_gateway_attachment_id
key = "Owner"
value = "Operations" }
```

For a certain provider, you can even give the same tags to all resources. This should be handled with care as you don't want to wrongly tag other resources, especially if those trigger third-party integrations or cause wrong billing:

```
provider "aws" { # ... other configuration ...
default_tags {
tags = {
Environment = "Production"
Owner = "Ops" } } }
```

Practically all IaC solutions provide this approach and can be adapted to any kind of infrastructure, cloud or on-prem.

This can be done for Kubernetes resources, too. We will skip regular deployment files here, as those will most likely be handled differently, as with Helm. The following example shows how Helm takes the value for the labels from the Helm chart value file. Template values are dynamically created by a `_helper.tpl` templating file, while `.Release` values are built-in information. Another option is to read the values from a `values.yaml` file, which is provided by the user:

```
apiVersion: apps/v1
kind: Deployment
metadata:
 name: {{template "grafana.fullname".}}
 labels:
   app: {{template "grafana.fullname".}}
   chart: {{template "grafana.chart".}}
   release: {{.Release.Name}}
   heritage: {{.Release.Service}}
```

One strength of Helm is that we can combine this with a simple `if` statement. Through that, we can hand over a predefined set of information depending on where the deployment is going to be or based on any other trigger.

> **Note**
>
> Utilize predefined labels in your Backstage templates to force your users to have some bare minimum of labels defined.

Both approaches are heavily declarative and can fail in their adoption or being overseen. For sure, you can check the correct usage of labels in the CI/CD pipelines, but this might result in many deployments failing. This leads us to policy engines. On the one side, those can be used to test the deployments for labels, but a tool such as Kyverno can also add this information when needed. The following example of a Kyverno policy adds `foo=bar` labels to any Pod or service:

```
apiVersion: kyverno.io/v1
kind: ClusterPolicy
metadata:
 name: add-labels
 annotations:
   policies.kyverno.io/title: Add Labels
   ...
   policies.kyverno.io/subject: Label
spec:
 rules:
 - name: add-labels
   match:
     any:
```

```
    - resources:
        kinds:
        - Pod
        - Service
  mutate:
    patchStrategicMerge:
      metadata:
        labels:
          foo: bar
```

We can select and patch that information in a very fine-grained way, and it can be updated anytime later if needed.

Consolidated versus separated cost and billing reports

Within many organizations, we see the demand for consolidated cost and billing reports. While that is, for the organization, on a higher level, the fastest way to understand the spending on infrastructure, it also requires many tags to ensure a proper split and level of detail. When organizations are growing too large, and the amount of business tags is larger than the technical tags, it is a clear sign that something went wrong. Being radical at this point, it then becomes an anti-pattern to do consolidated cost and billing reports. But what can we do, then? Enterprises, for example, often work with alphanumeric codes to identify departments. So, instead of a department being called *Finance and Accounting*, they may have something like B-2-FA-4. Such code is the perfect groundwork to eliminate many other organizational or business tags and place them into a matching table or database, in case you want to programmatically do this matching somewhere else. Other things to consider are, for example, the staging environments and the cluster. Most of the time, development, integration, and production systems are standalone. This means if your app is running on the Manhattan cluster, which could be translated to prod-1234-eu, then I don't need another tag that says stage=production or region=eu. Just to be clear, you should limit yourself to not adding too much information into a single value.

However, there is a big *but* in this story that leads us to the beginning of the topic. From a cost management perspective, you need to have many good tags so that you are able to do your analytics and research.

Tags themself are not enough; they are a relevant part of the cost optimization, but which optimizations do you take? In the next section, we will look into general approaches for optimization strategies that are not only rightsizing and reducing infrastructure.

Looking at cost optimization strategies

The fastest approach to reduce cloud and platform costs is through shutting down what you don't need. The biggest problem is that everyone involved in that process might have some reason why they require the infrastructure as it is. But I would like to turn your perspective on this topic from here onward: *Cost optimization is nothing that we should introduce afterward.* Here's yet another principle you should consider: *Be cost-aware and effective, considering it in the design of the platform.*

> **Note**
> The more cost optimization potential can be found in a platform, the worse we have done in our job as platform engineers and architects.

We can leverage potential on any part of our platform. However, this could be a book in itself to cover all aspects of it. Therefore, we will cover principles that are applicable to any component within the platform.

Streamlining processes

Processes, business and technical, can cause higher costs. I will give you an example of a pattern/anti-pattern that leads to unnecessary costs. The Kubernetes release is divided into three major releases, and every few weeks in between, you will get patches, bug fixes, and security patches. The Kubernetes release team automated a huge part of the release process, having nightly builds and running *all the time* several thousand tests on the Kubernetes infrastructure. When this approach was introduced, the community took it as the holy grail of the top tech companies. Suddenly everyone wanted to have nightly builds of their containers, often skipping the test part, which would be too much effort.

The downside of this approach is that most nightly builds never get deployed. It also prevents build servers from shutting down at night. The container registry is kept busy, including CVE scans, and the enterprise containers are usually more on the heavy side, raising transfer and storage costs. What most organizations didn't understand was that Kubernetes is a global project. Also, if for someone it is night, somewhere else, there are people up and working on new features and pushing them to the Kubernetes repository. In the end, Kubernetes is one of the largest open source projects, initiated by Google, the largest digital-native company.

> **Think it through!**
> Just because everyone else does it, think twice as to whether it really makes sense!

Back to processes. Almost every process can be improved, as they are usually grown historically. Within the platform space, regular tools are introduced that replace scripts and own custom developed tooling. Better processes not only cut costs but can increase efficiency, reduce risks, minimize errors, and provide consistency.

Proven approaches for process optimization can be found in the *Lean* and *Six Sigma* methodologies. To improve a process, they teach you to do the following:

1. Define/identify the problem of the process.
2. Measure the performance and time of the transaction.
3. Analyze the performance for inefficiencies and dependencies.
4. Improve the process to solve inefficiencies.
5. Control and review the new process by introducing canary deployments.

The fun part of platforms is that most processes can be found within CI/CD pipelines and GitOps implementations. Those translate our target into technical steps to ensure the desired state. Where it becomes complicated is when we have process dependencies within Kubernetes and components on top of it. As an eventually consistent event-driven system, it is a challenge to identify which dependencies are causing which behavior. Sometimes, we cannot change this component reaction without causing further impact.

Processes in a technical world often sound like a bunch of scripts playing together. An ideal state is a cloud-native approach that utilizes the Kubernetes standardized API, controller capabilities, and resource definitions. It decouples different capabilities, and with those processes, makes them easier to optimize. Also, it makes it difficult to introduce new, historically grown process trash.

Finding the best deals for the best prices

A straightforward way to optimize costs is to compare list prices and take the cheapest option. *Cheap* is thereby relative because you might cut down the performance, the throughput, the available IP addresses, and so on. Costs and utilization are good indicators but have to be evaluated for their second-tier impact. Reducing, for example, the instance size might reduce the throughput and performance, which causes you to have to introduce another instance. That way, you might utilize two instances for 80%-90% but you still pay the same or even more than if you ran a larger instance with fewer utilizations.

If you are not bound to a region, some regions are cheaper than others; even so, it has no large effect. A strong option at the moment is the utilization of ARM servers. If you are not able to run your own or users' workload on it, at least introduce it for managed services such as databases. This can cut 20-30% or more of costs.

Those are very obvious steps. Find and identify the right resource that is not overprovisioned in the right region. And, as explained earlier, often you can achieve better cost performance when you keep things lean and clean.

There are many tools, such as Cloudability or Flexera, that can help you identify a cheaper option, and some even recommend changing architectures to reduce costs. These tools are extremely helpful to get started, but they often also come with a high price, and the knowledge coded into them is nothing you couldn't gain through some research by yourself or attending free cost optimization classes.

Designing for the highest utilization and lowest demands

As platform engineers cut down the infrastructure, it will most likely have an impact on the user experience. The more flexible your platform becomes, the better it can adjust to the given situation. You can introduce active components that declutter and reduce the size of the system where possible and educate your user about the possibilities to increase and decrease the workload as needed. Matching their utilization with the bill can motivate the user to take care of responsible usage.

Previously, you learned about the difficulties between many small nodes and a few larger ones. We also discovered that we can easily support different CPU architectures and allocate resources dynamically. All those elements play a role in the definition of the core of the platform. I always have in mind the picture of an ant colony. At their center, they have a lot of action going on, but the core is a stable construction containing the majority of ants in it. When the colony, aka our platform, grows, we will extend this part. But sometimes, we don't know if it isn't just temporary. Throughout the book, you have seen different ways to handle dynamic workloads, and in the following section, we show some best practices for scaling.

So, what would be an ideal picture? Besides the aspect that it depends on your demands, the platform should utilize its resources as much as possible while having no overhead infrastructure. Simply said, it's not easy to do as a platform. The workload is not within your hands and can be anything from static long-running resource-intensive software to thousands of serverless containers being continuously on the move. What we can take from this is that you can design the platform as well as you want; in the end, it all comes down to a continuous job of analyzing, reacting, and adjusting. With time, you can build a base layer of automatic reaction into the platform that will cover most cases, but you still will be required to work on the optimization.

In the last part of this chapter, we take a look at some concrete examples of scaling and optimization. You will learn about reactive and predictive scaling as well as cost-aware engineering.

Autoscaling, cold storage, and other tricks for cost optimization

In *Chapter 4*, in the *Autoscaling clusters and workloads* section, we already discussed a key benefit and core capability that comes with Kubernetes. Several different tools and mechanisms are built into K8s to scale, such as configuring ReplicaSets (the number of instances per workload we want) or using observability data to drive automated scaling decisions using **Horizontal Pod Autoscaler (HPA)**, **Vertical Pod Autoscaler (VPA)**, or **Kubernetes Event Driven Autoscaling (KEDA)**. There is a great free tutorial that walks through all the different options to autoscale on Kubernetes provided by *Is It Observable*. Here is the YouTube tutorial link, which also contains links to the GitHub tutorial: `https://www.youtube.com/watch?v=qMP6tbKioLI`.

The primary use case of autoscaling is to ensure that workloads have enough compute, memory, and storage to achieve certain availability goals. For our Financial One ACME company, this could mean that they use autoscaling to ensure that their financial transaction backend can handle 1,000 concurrent transactions to be processed within a 100 ms response time. While autoscaling can help us reach our availability objectives, it also comes with a price tag as scaling resources means somebody needs to pay for that extra compute (CPU) or memory. Improper autoscaling—scaling too much, never scaling down, or scaling at the wrong times—can also lead to unplanned cost explosions while not meeting the real objective of autoscaling!

Autoscaling done right is what we want. Doing it right not only allows us to achieve our business and technical objectives but also allows us to leverage autoscaling to keep costs under control. Let's have a look into several autoscaling topics platform engineers should be aware of and what else we can do to optimize costs. Be reminded that some of the practices we will discuss in the following sections are also applicable to any type of workload: cloud or not.

Many shades of autoscaling

The scalable nature of the cloud and Kubernetes as a core platform is a great capability to have at hand as platform engineers. There are, however, many different ways to scale. There are different trigger points that can cause a system to scale, plus, scaling must not just go in one direction (typically up) but we also have to consider scaling systems down—even all the way to zero to avoid wasting resources for nothing!

Scaling up – not only on CPU and memory

Most engineers are familiar with scaling based on CPU and memory. Most examples you will find for HPA typically scale the replicas of Pods based on a certain average CPU utilization of that Pod. That's the most common way of what we also refer to as reactive scaling as we scale based on reacting to a reached threshold.

However, scaling based on CPU or memory is not always the best option. It's also not the only one we have, even though most scaling frameworks initially started with just scaling based on CPU and memory as those are the two key limits one can set on Pods or namespaces.

For services that are optimized for a certain throughput, it would, for instance, make more sense to scale on concurrent incoming requests. Also, for the CPU, instead of scaling based on average CPU utilization, it could be better to scale your system when K8s starts to throttle the CPU for a Pod. In our previous chapter, we mentioned tools such as KEDA as well as the **Cloud Native Computing Foundation** (**CNCF**) Keptn project, which can provide any type of metric from various observability sources (Prometheus, Dynatrace, Datadog, New Relic) to be used for event-driven autoscaling. To see how this works, check out a full example in the Keptn documentation: `https://keptn.sh/stable/docs/use-cases/keda`.

The key takeaway is that not every workload is CPU- or memory-bound, and Kubernetes is not limiting you to just defining your scaling rules based on those two key attributes. Defining scaling rules based on what is really allowing your workload to execute more efficiently will also lead to more efficient resource utilization, which leads to a system that is also optimized for costs!

Predictive versus reactive scaling

It is often assumed that autoscaling works instantly. But that's not true! Think about Financial One ACME. If on paycheck day everyone wants to check their new account balance, this might mean a 10x spike in regular traffic within the first couple of hours of that day. The cloud vendor, however, cannot guarantee that all those resources are instantly available as you are competing with many other organizations that are also trying to request cloud resources at the same time. On top of that, the workload itself will not be able to process incoming requests instantly as many of those Pods have a certain startup time until they are ready.

This problem can be solved through predictive scaling. Compared to reactive scaling – which is scaling when we reach a certain threshold, as explained in the previous section – predictive scaling looks into a potential future situation and reacts to it before reaching a threshold. Predictive doesn't mean we need a magic glass bowl to tell us the future. It can be as simple as starting the scaling a couple of hours before we anticipate a spike in traffic; for example, just ahead of paycheck day or ahead of a marketing campaign that starts at a specific time.

Other predictions can be more dynamic:

- **Seasonality**: One could be based on seasonality by looking at historical data. E-commerce is a good example where, throughout the year, you have certain dates that see a spike in load, such as Black Friday or Cyber Monday. It's easy to *predict* those spikes!

- **Related data sources**: Another one could be to look into other data sources. Insurance organizations often look into severe weather data. When storms are predicted and there is a certain chance that those storms will cause damage, it makes sense to predictively scale services that those clients will use to submit insurance claims. For this particular scenario, you can even scale in specific regions that are closest to the storm.

- **System dependencies**: In complex systems, it is also an option to scale dependent components based on the load behavior of other parts of the system. Take hospitality as an example. If we see more people searching for flights as they want to get away on a long weekend, we may also predictively scale backend services that provide recommendations for hotels, cars, or additional events to book in the travel destination.

Use case – predictive storage scaling to optimize cost and availability

Now that we have learned about different approaches to predictive scaling, let's apply this to a very costly example: storage!

Our digital systems are generating more data than ever, and it's predicted that this trend will continue. Storage – while seemingly available in abundance – is a big cost factor for many organizations. The same is true for our Financial One ACME company, assuming we have to store all the details about every financial transaction our systems handle. As an organization, we need to make sure we can always persist all records, but we also want to make sure that we are not paying for storage space we currently don't need. So, we want to keep our free disk space as small as possible to avoid having to pay for disks we don't need. On the other hand, we also need to make sure we have enough disk free space in case a spike in transactions comes in as we cannot afford to be able to lose transactions. Considering that scaling disks at a large scale can't happen instantly but may take a couple of hours, we can apply predictive storage scaling to fulfill all our requirements!

The following screenshot helps us understand how this works. It shows the disk free space % metric over time. The more data we write, the less disk free space is available. Instead of scaling our cloud storage on a certain fixed threshold, we can use a predictive model and scale when we predict that we will reach a certain low threshold within the timeframe it takes to scale – hence ensuring that we always have enough free disk space without paying for too much:

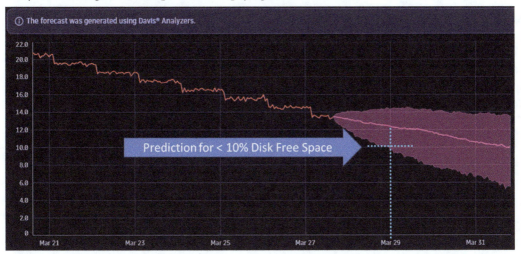

Figure 8.4: Predictive scaling of cloud storage to optimize for cost and availability

The preceding example is taken from a real use case that resulted in great cost savings by continuously rightsizing storage using a predictive scaling approach!

Scaling to zero – shutting things down when not needed

While many systems we operate need to be available 24/7, some systems don't have that requirement. These might be systems that are only used by employees during regular business hours or systems that are only needed to fulfill certain jobs during special times of the day, month, or year. I am sure we can all think about systems that are often sitting idle yet consuming precious resources and therefore causing costs even though they are currently not needed.

Virtual machines (**VMs**) are great examples of scaling to zero. We have been doing this for many years to take snapshots, shut VMs down when not needed, and bring them up again when the work had to be continued. Many organizations fully automated this by automatically shutting down VMs that were used for daily business tasks at the end of the business day and bringing them up again the next morning. This alone is a great cost-saving opportunity as many VMs can be shut down during nights and weekends!

In Kubernetes, we also have the opportunity to scale workloads to zero. We could use the already discussed KEDA but also look into tools such as Knative (which can run serverless workloads) or `kube-green`. The latter was created to reduce the CO_2 footprint of K8s workloads and clusters and is able to put workloads, nodes, or clusters to sleep when not needed. To learn more about `kube-green`, check out the following website: `https://kube-green.dev/`.

A question we still need to answer is: Which workloads can be scaled to zero and for how long? We get this data from the owners of those workloads by specifying when and how long they need them. Another approach to this is simply using observability data to see which workloads are used at which times during the day and, based on that, create a `kube-green` sleep configuration to scale workloads to zero. An example of this implementation can be found in *The Sustainability workshop* from Henrik Rexed: `https://github.com/henrikrexed/Sustainability-workshop`.

From scaling workloads to clusters

So far, we talked a lot about rightsizing or scaling workloads to ensure we have enough resources to meet our business objectives but also to not overprovision so that we can optimize on cost.

As we scale our workloads up and down, the underlying Kubernetes cluster also needs to be sized accordingly. This is where cluster autoscaling comes in, which will scale up a cluster's nodes to ensure that enough resources are available to run all workloads, but which will also scale down nodes in case nodes are underutilized and workloads can be distributed across the remaining set of nodes. This ensures that the underlying cluster node machines are optimized, which in the end saves costs.

There is a lot of existing documentation on the Kubernetes documentation website: `https://kubernetes.io/docs/concepts/cluster-administration/cluster-autoscaling/`.

There are specific autoscalers such as Karpenter – initially developed by AWS – that help right-scale a Kubernetes cluster but also keep costs in check. Karpenter integrates with the APIs of your cloud vendor and is able to provision the right size of nodes that are needed to handle a certain workload. It will also scale down nodes if no longer needed.

In addition to tools such as `kube-green` (mentioned earlier), Karpenter is a great option to scale your clusters while keeping costs in mind. To learn more about Karpenter, check out `https://karpenter.sh/`.

As platform engineers, it's important to be aware of all the different scaling options. Many of those can be set up to rightsize workloads and clusters. For some, it's important to work closely with the engineering teams and workload owners to define scaling strategies that make sense for those specific workloads. Overall, autoscaling – whether it is compute, memory, or storage – is one of the key enablers of cost-efficient platform engineering!

Cost-aware engineering

Now that we have learned what we can build into our platforms to rightsize and autoscale to save costs, we also need to talk about what we can do to enable engineering teams to become more cost-aware from the start. The best cost optimization starts when anyone in an organization is aware of the cost impact their actions have and therefore starts building and designing systems that are more cost-efficient by default. Reporting on costs based on tagging is one way of making teams aware of their costs. This strategy has been discussed earlier in this chapter.

Let's look into some additional options we think everyone should consider as they have the power to lead to cost-aware engineering!

The only request what you need approach

In the early days of the cloud, many organizations gave their engineering teams full access to cloud portals. This easy "self-service" boosted productivity as everyone could easily stand up new VMs, create new storage services, or even create Kubernetes clusters. This "wild wild west" approach, however, led to cost explosions for many organizations as users were just creating new services but not thinking about the basics, such as: How large a virtual environment do I really need and for how long do I need it?

One of the organizations that the authors have worked with is a financial organization. Instead of giving everyone full access to their cloud portals, they built their own self-service portal allowing engineering teams to create new VMs, databases, clusters, and so on. As part of that self-service portal, the team had to define for *which application* they needed the resource and for *which environment*, as well as *how long* this machine was needed; for example, only during business hours. The result was a 60% cost reduction as the provisioned services were automatically shut down when no longer needed. The following screenshot shows how engineers request resources for the time they need them. On the right, you also see the detailed reporting and overall optimization goal this organization is achieving:

Figure 8.5: A self-service platform that reports and reduces costs

The reporting in this use case was not only done on costs but also on carbon impact, which is a big topic for most organizations these days. Providing this self-service through a central platform made it easy for this organization to make their engineers more cost-aware from the start and also show them the positive impact their actions have.

Lease versus flat-rate resource

The previous example is great but requires teams to think upfront about exactly how long and when they need certain resources. A different approach to this would be to use a **lease model**. What does this mean?

When dev team A requests a resource – let's say a Kubernetes cluster that they need for some development work – they can simply request it for a default time period; for example, 1 week. That 1-week timeframe becomes their "initial lease" of that resource. Both timeframe and team ownership will be managed through tags on that created resource. Through automation, 1 day prior to the end of the lease, emails or chat messages can be sent to teams reminding them that their lease is about to expire. The message can also give them the option to *extend the lease* for another day or week but also remind them about the cost that this extra time would incur.

This approach has been implemented in several organizations and ensures that any team can still get the resources they need via self-service. It also ensures that resources that are forgotten about or no longer needed will be shut down without having to specify upfront exactly how long a resource will be needed.

Now that we've talked about how to have resources running only when they are really needed to save on costs, let's talk about how engineers can also optimize their code to make a cost impact!

Green engineering – optimizing your code

Efficient code typically not only executes faster; it typically also needs less CPU, memory, and potentially less disk storage or network bandwidth. Less of everything also means less costs. So, why doesn't everyone just produce efficient code from the start?

Too often, engineers are under time pressure to deliver new features, or organizations haven't invested in tools that test for and provide optimization recommendations as part of the software delivery life cycle. The authors have worked with many software organizations in the past and have identified a handful of very common patterns that result in inefficient and, therefore, costly code. Here are some examples:

- **Requesting too much data**: Instead of leveraging query languages to only request data needed for a certain operation, more data is retrieved and then filtered and processed in memory. This leads to more network traffic to transfer the data, more memory usage to store the data, and more CPU to iterate and filter in the client code.

- **Inefficient use of libraries or algorithms**: Many software libraries exist to get certain jobs done; for example, **Object Relational Mappers** (**ORMs**) to map data in databases to objects in a development language. Development teams unfortunately don't always have the time to properly test or configure those libraries to optimize them for their specific use case. This, therefore, results in inefficient use, which leads to higher CPU, memory, network, and disk access.

- **Excessive logging**: Software engineers use logging frameworks to log information as their code executes. Logs are typically needed for analytics, diagnostics, and troubleshooting of failed or problematic code executions. Too often, though, logs are created excessively or duplicated without proper formatting or enough contextual information; for example, no log levels are set. This leads to overhead when logs are created but also overhead when those logs are ingested, transformed, and analyzed by observability platforms.

There are many more patterns in software engineering that lead to performance or scalability issues. Apart from detecting patterns, architectural reviews for applications can also lead to cost reduction through more efficient architectures or rewritten code. A prominent example is Amazon Prime Video, which dumped its AWS distributed serverless architecture and moved to what is described as a *monolith* for its video quality analysis, which reduced infrastructure costs by 90% *[4]*. Ultimately, those patterns also mean inefficient code execution, which results in higher costs. As platform engineering teams, we have the opportunity to analyze those patterns using modern observability tools and bring this information back to the engineers to remind them not only about the costs they incur with their code but also where they can start optimizing, as shown in the next screenshot. These two charts show how many logs are created per service and also highlight which logs do not have proper configurations; for example, no log levels set:

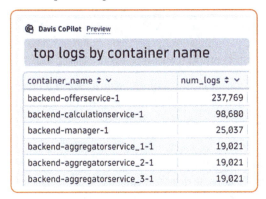 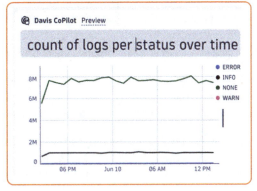

Figure 8.6: Providing teams with easy insights into patterns such as excessive logging

This brings me to the last part of this section, which is the opportunity to educate engineers and make them cost-aware from the first line of code they write!

The education opportunity

While this may not stand out as a top role of a platform engineering team, as engineering teams use our platforms to deploy their applications as self-service , we can use that platform to also educate everyone about the cost impact they have when using the platform to get their software services deployed.

In the previous sections, we already highlighted use cases such as sending cost and usage reports to engineering teams or identifying and highlighting inefficient code patterns. The key enabler to this is proper tagging (for example, who owns which part of the infrastructure and applications), as well as good observability (for example, which systems use how much CPU, memory, network, etc.). Having this information allows platform engineering teams to push this data to teams proactively and, with this, show them continuously what cost impact their applications have. Doing this continuously will also result in an educational effect that will lead to engineers having a better upfront understanding of the cost impact their actions have.

Summary

In this chapter, you should have developed a sense of costs and had ideas on how you can address this topic on your platform. Good platforms provide transparency for their user and enable the use of flexible options to adjust their workload for different triggers. At this point, you should be able to combine the learned approaches from previous chapters, such as dynamic resource allocation with GPUs, to create high utilization and optimal cost allocation.

Remember that the cost perspective alone isn't enough to reduce overall platform costs, as some reductions in server sizes might increase the demand on multiple small ones due to other limitations that cloud providers have. A tagging strategy builds the core for control and transparency. What sounds easy can end in many organizational discussions. To optimize your costs, you can also leverage other elements such as processes, and agree on long-term commitments to get better pricing deals.

Finally, we gave you some practical examples and best practices for your platform to include. We caught up on the different approaches of scaling and the difference between predictive and reactive scaling, and shone a light on other scaling factors besides CPU, such as memory and storage.

To summarize, when you think rationally and treat the money you spend on the platform like it is your own, then you can become very cost-efficient. As a platform engineering team, you could also develop a metric to define how efficient the platform is to be able to agree with your management on using this free budget for further investments and optimizations. Remember that even though the cloud gives us an *unlimited* amount of resources, we don't have to take all of them just because they are there.

Let us head straight to our final chapter. As we already said earlier, the only consistency is inconstancy. In our last chapter, we will therefore talk about continuous change and how to survive it, considering lightweight architectures motivated by sustainable ideas and the golden path for changes. To close the chapter, we dare to take a look into the crystal ball and cover some technological trends that may or may not become relevant within the next years.

Further reading

- [1] Why we left the cloud – David Heinemeier Hansson:

  - `https://world.hey.com/dhh/we-have-left-the-cloud-251760fb`

  - `https://world.hey.com/dhh/the-hardware-we-need-for-our-cloud-exit-has-arrived-99d66966`

- [2] FinOps Framework high-resolution poster: `https://www.finops.org/wp-content/uploads/2024/03/FinOps-Framework-Poster-v4.pdf`

- [3] Cloud Custodian: `https://cloudcustodian.io/`

- [4] Prime Video cost optimization: `https://www.thestack.technology/amazon-prime-video-microservices-monolith/`

Get This Book's PDF Version and Exclusive Extras

UNLOCK NOW

Scan the QR code (or go to `packtpub.com/unlock`). Search for this book by name, confirm the edition, and then follow the steps on the page.

Note: Keep your invoice handy. Purchases made directly from Packt don't require one.

9

Choosing Technical Debt to Unbreak Platforms

What is **technical debt** and how does it unbreak a platform? Technical debt is the ongoing cost of maintenance for a piece of software. This is the literal money that's spent on the runtime environment, the time spent operating and updating it, and customer satisfaction. Just like monetary debt, technical debt can pile up. As the amount snowballs, the team can't incur new debt or deliver new features as all their working hours are spent on mitigating the current issues.

Indicators that a team is inundated with technical debt are easy to spot: are your teams attempting to maintain extremely outdated software? Do you have systems that are operationally fragile and/or prone to crashing? Do you use unnecessarily complex software?

If you answer any of those questions with yes, there's a corresponding cost. For instance, updating, securing, and operating outdated software ranges from very costly to impractical. Operationally fragile software requires frequent manual intervention, so it will eat up your operations team's time and increase the chance of burnout. Last but not least, overcomplicated software is usually just plain hard to understand. It's hard to read and maintain, and this means ongoing development and maintenance costs.

No software can be built or adopted that is debt-free, but when that is evaluated alongside its features, it can help guarantee a more successful running state for the platform. In this chapter, we'll cover the following topics:

- Taking technical debts consciously

- Using data to drive design decisions

- Maintaining and reworking technical debts

- Rewriting versus refactoring – a practical guide

- Architectural decision records – document for the Afterworld

Taking technical debts consciously

"Choose your technical debt wisely," your author, Hilliary, says all the time. But what does it mean to choose your technical debt? Aren't we supposed to be solving problems and identifying solutions?

Yes. But these two things are not mutually exclusive. There's not likely to be a perfect solution readymade off the shelf. Typically, products are designed with specific use cases in mind, but those use cases are not all-encompassing. No product can cover every permutation of every user's potential needs and workflows. Indeed, it will be the same for your platform as you progress through its life cycle. When we commit to accepting the shortcomings of a solution or technology, that means we're committing to compensating for that in another way. Such compensation should be managed so that we don't create additional overhead or manual workflows that result in unsustainable toil for the platform team or the developers. Too much technical debt takes away from the return on investment in the platform.

Technical decisions should be made collaboratively. It's important to try to arrive at a consensus as a team to ensure that the path forward is one at least the majority feel comfortable with. To take on technical debt consciously, you must be able to assess it and understand the answers to important questions.

The following is a technical debt assessment for you to try out:

1. What problems are you solving today?
2. What problems do you not have (yet)?
3. What will this cost us in time to build or adopt?
4. What will this cost us to run?
5. What is necessary to maintain it?
6. What is our expected return?

7. Can the team sustain it *as-is*?

   - If not, is the team prepared to upskill and do they have time?

   - If it's open source software, how strong is the community? How does the team join that community?

8. What will it take for any new team members to get up to speed?

9. When will we know if it's time to scale?

10. How will we scale?

11. What are our deprecation criteria for this solution?

Visualize the answers to these questions as an accounting ledger. After you total everything up, are you coming out ahead or behind? Like your books, these values must be balanced. The balance of technical debt for a team or an organization is more socio-technical than technical. The artistry of engineering can lead to an emotional connection that may create blind spots and flaws that impact the long-term sustainability of a solution. For example, tools written by passionate engineers on the cutting edge more often fall into disuse and disrepair if they weren't written in a way that invited collaboration from the team, and if the technical pieces exclude a number of the team without the skills necessary to maintain those tools.

A solution can be as clever, elegant, and simple as possible and still be unsustainable if it's not suited for the team or the environment in which the problem being solved exists. For example, if the solution is written in Rust, and the team to date has been using Ruby, then it's unlikely the team will be able to sustain the solution if the person who designed and implemented it leaves.

That's why of all the questions, the last becomes the most important. Deprecation criteria tell us when it's time to retire or replace a tool, a system, or any piece of code. All things eventually become legacy software, so a deprecation plan or criteria against which you evaluate the notion of deprecating is an important aspect of sustaining technical debt.

Moving beyond the thinnest viable platform (TVP) sustainably

When we build a TVP, we commit to the notion of *good enough*, but as we evolve, what was good enough at one point may start becoming not right. This is the first iteration of the platform's technical debt. Technical debt is a sign of growth and evolution. It may sometimes be the result of having taken what in retrospect was the wrong decision, but it is just as often the result of moving beyond the point you were at when the decision made sense. To some degree, technical debt is the cost of doing business, while in other respects, it can be considered growing pains. However you look at it, it's a symptom that needs to be managed or mitigated.

When it's time for the platform to move beyond its thinnest and start on its path toward being the robust solution expected to sustain an organization at scale, we start to reevaluate our past decisions. Do we expand our existing features? Do we replace any components? Do we retire anything that didn't perform how we expected? This transition into the next generation of the IDP is led by the same data used to define the TVP.

Critical user journeys

Initially defined in the discovery or planning phase for your IDP, the **critical user journey** defines the expected workflow of interactions between a user and the IDP and where and how they expect to be successful.

However, as organizations evolve, so do these critical user journeys. It's possible for things that used to be very important to an organization to cease to be as important or to become of lesser importance to something else that has emerged over time.

Additionally, if you followed the TVP pattern, then it's likely that the initial phase of building out the platform excluded some use cases and potential critical user journeys. A regular review of these journeys and consciously deciding to add another or expand on an existing one will shape how your platform grows and how your team innovates.

A critical user journey is built around a user story. For *Financial One ACME*, there will be a need to cater to a certain geographical region, so a user journey would be phrased in the following way: "*As a user of the platform, I want to deploy an application only accessible from North America.*" Once a user story has been accepted by the platform team, the user journey should be created. Not every proposed user story should be accepted right away. For example, in the preceding story, the regional boundary to access requires supporting infrastructure. If that use case is not business-critical, could be satisfied without additional work, even if sub-optimally, or can be supported outside of the platform, then a different user story may be more important to tackle first.

Assuming the user journey "*Deploy an application only accessible from North America*" is accepted, we could implement it using tools such as Crossplane and the composite feature, which allows the platform engineering team to define a template to deploy an application in a certain region. The user in their journey then simply needs to create an instance of that template, specify all the values, and commit it to Git. The rest is taken care of by the platform's core delivery tools. The result will be the application that gets deployed, which is only accessible from the specified region. Technically, this can be done through specific ingress routing rules for regional-specific domain names, as shown in the following example definition:

```
apiVersion: composites.financialone.acme/v1alpha1
kind: FinancialBackend
metadata:
  name: tenantABC-us
spec:
  service-versions:
```

```
    fund-transfer: 2.34.3
    account-info: 1.17.0
redis-cache:
    version: 7.4.2
    name: transfer-cache
    size: medium
database:
    size: large
    name: accounts-db
region: "us-east"
ingress:
    url: "https://tenantABC.financialone-us.acme"
```

For more details on Crossplane, please revisit the *Workload and application life cycle orchestration* section of *Chapter 4*.

Evaluating the importance and the value of each user story helps to grow the platform sustainably. The team won't be able to do everything all at once, so it's important to keep the scope from creeping up at a rate the team cannot sustain.

This is the same principle that should be applied to all software engineering, and in this way, the platform and an end-user-facing application are no different.

Avoid over-engineering

What is over-engineering? **Over-engineering** is typically characterized by making a very robust and fully featured system that tries to solve every conceivable problem, even before the problem exists. Over-engineering is an easy but dangerous trap to fall into. The first two questions in our assessment help us to avoid over-engineering solutions. Essentially, we're using those questions to define our goals and our non-goals for each decision. Understanding what problems and use cases we're solving for and defining what we're not solving for – or at least not yet – can help guide our decisions and keep our project scope sane.

For example, in an observability solution, it's reasonable to expect a lot of data over time. It's easy to fall into the antipattern of planning for infinite data and building infrastructure accordingly. Instead, ask targeted questions:

- How much data is really useful?
- For how long is it worth it for us to keep this data?
- How long *must* we keep the data?
- Who needs access to the data?

By defining parameters based on your answers, you will allow the data to remain manageable. If there isn't a use case for keeping data beyond a month or a year, then the solution shouldn't be to write a system that can handle huge amounts of data. A better solution is a system that handles small amounts of data and regularly purges what it doesn't need. Alternatively, if full purges don't meet your use cases or feel too extreme, then data aggregation may be the correct approach. For example, telemetry data that's collected over several seconds can be rolled up into days, weeks, months, or even years. Data aggregation can allow for data that covers a long period to still exist but without the same need to engineer storage and retrieval mechanisms for vast amounts of data. If you're feeling like you're having flashbacks to *Chapter 7*, well spotted. The same questions that help prevent over-engineering also help us to understand our security posture in many cases.

Avoiding over-engineering doesn't mean never planning for the problems you don't have yet. Failure to plan for the future can also lead to creating solutions that may need to be replaced entirely down the line. Now, this can be good when we look at composable system architectures and the power of microservices. Sometimes, a complete swap of a component is the best way to get the best results. However, it's not usually an easy or quick thing to do, so if that's what's desired, it still needs to be carefully planned and executed along that plan. A general recommendation to avoid overengineering is to solve today's problems and roadmap tomorrow's. Keeping systems simpler allows them to be more easily maintained and can help ensure more flexibility as the product evolves.

Build versus buy – building a decision tree

In the process of adding a new component and evaluating technical debt, one of the most important questions is whether we should build it or buy it. This tends to be more sociotechnical than technical as it involves balancing the innate desire to build with the need to buy if that's what's right for the team.

It's natural for engineers to want to engineer things. Our identities, strengths, experiences, and interests bias us toward decisions that align with them. However, it's critical to take a step back from the emotional and intellectual attachment to solving the problem and look at the options with as unbiased an eye as we can.

The following figure is of a decision tree on whether or not to build or buy a solution. Please note that *buy* doesn't always mean spending money; it may also refer to the adoption of a free or open source tool:

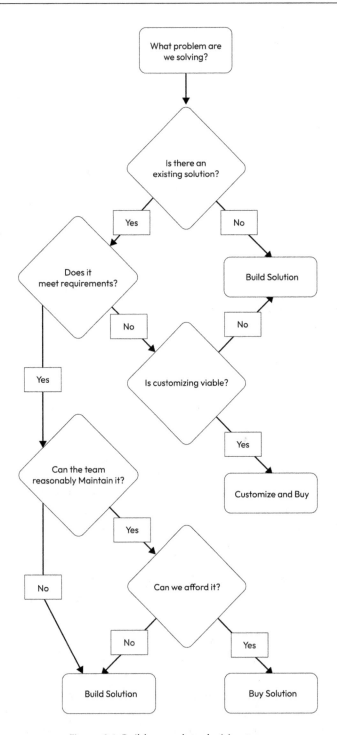

Figure 9.1: Build versus buy decision tree

Following this decision tree should be done as a group. Some of the questions may have answers that are more subjective or that different team members have differing opinions on. Collaboration is key for moving these decisions sustainably.

Let's work out an example decision tree together. In this case, our fictitious company *Financial One ACME* is looking to modernize an older technology stack. As they aim to do this, they'll likely need to address technical challenges that didn't exist for them previously, or look to integrate new functionality that wouldn't have been possible in a system that didn't use more modern architectures.

One such example of new or different functionality might be assessing the adoption of data transformation tools. With more data capabilities existing in a cloud-native architecture, either the product team, the platform team, or both might consider utilizing such a tool to make raw data more useful.

For example, if the platform team wanted to collect usage metrics for the platform with Prometheus and help turn that into DORA metrics, a data transformation would be useful in that workflow. Similarly, other use cases for data transformation could be useful across the company, so the implementation would need to take several user needs into account.

If we oversimplify this example, we would follow the decision tree as such:

Q: What problem are we trying to solve?

A: We need to enable automated data transformations.

Q: Is there an existing solution?

A: Yes – Apache Airflow and Argo Workflows.

Note that in this case, the team has identified two potential solutions. From there, the next set of questions should be conducted for both solutions. Does it meet the requirements? Can the team maintain it? And can we afford it? Since both solutions are open source, the affordability question would be more about comparative costs to run, which may not be easily discernable up front. In assessing open source, the most important question when a tool meets the requirements is the team's ability to maintain the solution. The capabilities of the team can't be answered in a vacuum and require the team to weigh in on their capabilities and comfort for a new integration. If upskilling is required, that doesn't mean that the technical debt isn't reasonable to take on, but that time must be added to the schedule for everyone to get up to speed so that an answer of *This is potentially sustainable* doesn't become *This is unsustainable*.

The criticality of team buy-in

While many platform architects will be involved in the day-to-day running of an IDP, it's equally likely that they will not be. As such, the folks who will be hands-on keyboard managing and maintaining the day-to-day need to be fully invested in the direction of the IDP and its offerings. This will often be the platform team, but if you've been building a platform that encourages self-service and collaboration with the entire engineering organization at large, the team will be much larger in scope.

There may be cases where it's impossible to gain 100% consensus, at which stage it will be necessary to *disagree* and *commit*. Team consensus can reduce burnout, but that must be balanced with the need to ensure the IDP continues to evolve along its roadmap.

Team buy-in is incredibly important in decisions that will require the team to upskill in some way, such as learning the tool itself or a new programming language. If the team is unwilling or unable to adapt to the technological solution, the solution will not provide value in the end, regardless of its other redeeming qualities.

Now that we're familiar with the concept of technical debt, we can investigate mitigation strategies for that debt by leveraging data.

Using data to drive design decisions

So far, we've discussed the importance of observability data for self-service and developer satisfaction, but not all other use cases for the telemetry data an IDP can collect. Data is a powerful asset that can be used to aid decision-making and help in cost and time savings. For example, when looking at two solutions that seem to split the middle on solution fit and technical debt, it can be hard to know which direction to go. This is where you can leverage data to make your decisions, and to keep track of your technical debt.

Nearly all the data your team will need to leverage for decisions will either already be in your observability solution or be an easy addition to it.

Observability is key

After spending all the time and effort to set up an observability stack and collect and retain data, it would be almost irresponsible to only use that data for incident management and response. Fortunately, platform teams are rarely accused of being irresponsible, so it's likely you already know that this data can be used in more meaningful ways. Applying observability data to decision-making and using it to identify and manage technical debt are key use cases for your observability stack that are sometimes overlooked or under-emphasized. In *Chapter 6*, we talked about where and how that observability will be applied, but let's look at those same aspects now for technical debt.

Operational data and toil

We briefly introduced the concept of a **service-level objective** (**SLO**) in *Chapter 3*. We looked at it again in *Chapter 6*, as a key factor in ensuring customer happiness. However, SLO data can also be used to examine and identify technical debt.

A SLO is a target. Rarely is that target defined in the first round at the most ideal state since it can be impossible to achieve that out of the gate on a new project. A good SLO definition for a new project will be far more conservative but have a roadmap to the final desired target. Good SLOs are SLOs that evolve with the system they measure.

For example, if the ideal SLO is that 99% of application deployments over 28 days will be successful, then an initial target may be as low as 60%. In a situation where everything is greenfield, it's reasonable to start at a much higher target, but if we look at *Financial One ACME*, which is a company that is migrating a significant legacy offering, you know that the team there is starting in a disadvantaged position. Therefore, setting reasonably achievable targets and defining the next after those have been met is a more sustainable approach. This allows the team to stack rank the issues that are impeding success and demonstrate wins. You may be asking yourself, *"Is this gamification?"* *Yes. Yes, it is*. However, positive feedback loops are proven effective for reducing burnout and increasing developer happiness. Teams should be set up to win, not fail.

SLOs are not the only operational data that matters, however. It's entirely possible to maintain an SLO through manual intervention and not automation. This is where reliability engineers typically start discussing the concept of toil. Toil is a boring, repetitive task that must be done regularly or manually. Even a partially automated process is still toil if it requires a human to push the button to get it started.

Some toil is acceptable, such as manually starting a release pipeline for an organization that's not ready for a full CI/CD implementation. But other toil is not, such as an engineer executing the release for each component of the software every time an upgrade is necessary. Since toil is done by humans, it can be harder to track since tracking toil is toil itself. However, to manage technical debt, maintaining an awareness of toil and creating an action plan to reduce it is critical for ensuring the team has an acceptable level of cognitive load and a continued ability to innovate.

Toil for the team and the developers must be looked at both on a case-by-case basis and holistically to avoid death by a thousand papercuts. Just as we interrogate each piece of data we store or each new component we add, every manual task must meet the same scrutiny:

- Why do we need to do this?
- How long does it take?
- How long do we need to do it?
- Can we automate it?

If the answer to *"Can we automate it?"* is *"Yes,"* then that must be immediately prioritized and ranked in the backlog of team items, preferably highly since developer time is one of the most valuable assets of the organization aside from data. Each team's tolerance for toil will vary as it can depend on the composition of the team and their skills. The *Google SRE* (`https://sre.google/sre-book/table-of-contents/`) book defines a target maximum for toil at 50%, but teams are going to have their own tolerances based on team personality and other organizational factors.

Let's look at an extremely time-consuming task that needs to be done only once. Imagine one you've done before; it doesn't have to be anything in particular. Some engineers will see the manual task and will just get it out of the way. Others will assess the task, and they'll automate it even if it's a one-off and even if that makes the process take longer.

This might be because automation that's been peer-reviewed feels safer to use than steps that can risk human error or because they want to never have to do it again and they'd rather not risk being wrong about the task being a one-off task. This may also be simply because codifying what was done ensures it's not lost to time and tribal knowledge. Both doing the task and creating the automation for it were examples of toil, but despite taking longer, the process of automation handles technical debt better.

Finding your actual tolerance for toil will be a case of trial and error, but one tactic for keeping on top of toil is to set a SLO against it. By treating time spent on toil as a SLO, when that SLO is breached because, say, no more than 50% of our time is spent on toil, then a root cause should be conducted to determine why more time is/was spent and identify the corrective actions to get toil back within expected boundaries. However, teams should feel safe in doing things correctly, even if that means the error budget for toil is burned for one reporting period.

DORA metrics

If we look back on the DORA metrics and ratings we defined in *Chapter 2*, it's easy to see how these metrics can help you remain aware of your technical debt:

	Elite	High	Medium	Low
Deployment Frequency	On-demand	Between once a day and once a week	Between once a week and once per month	Between once per week and once per month
Change Lead Time	Less than 1 day	Between 1 day and 1 week	Between 1 week and 1 month	Between 1 week and 1 month
Change Failure Rate	5%	10%	15%	64%
Failed Deployment Recovery Time	Less than 1 hour	Less than 1 day	Between 1 day and 1 week	Between 1 month and 6 months

Table 9.1: DORA metrics competencies

If you've implemented something such as **Keptn** or another DORA metrics gathering solution within your IDP, then you've likely started tracking these metrics on every component within your IDP. As you track these metrics over time, you may notice certain trends.

For example, if lead time to change or deployment failure rates are still in the low to medium range after 6 months, then it would be worth investigating the reasons behind that to see if there's unaddressed technical debt.

Application performance data

Relevant to DORA metrics but adjacent in other ways is the performance data for the applications and components within an IDP. While the platform team isn't responsible for end user applications, the platform knows the state of the pods and containers the application uses. It can also know the success of API calls and response times. The platform team must enable the developers they serve to understand that state as well so that they can evaluate those metrics and determine if something is amiss. This will allow developers to manage their technical debt. This is a key aspect of the serviceability and self-service of the platform for the ongoing success of the company. Remember that the platform, once in use, is the most critical application within the company; it needs to support the success of itself and the end-user-facing application equally.

An application may be architected in a highly available way, fully leveraging the power of the K8s platform, but this could also allow it to unintentionally disguise problems or patterns that may be concerning. As such, application performance should be measured and surfaced regularly. As an example, in a system designed for high availability, a pod with three *ReplicaSets* may allow the application to run uninterrupted or with the appearance of no interruptions, even if there's an occasional OOM or other crash. However, the platform would know about those crashes and observability can be used to ensure that the developers who own the application are aware as well.

In *Chapter 6*, we discussed ways for the platform to support observability, be a little prescriptive where necessary, and allow developer self-service. Let's look at an example of where the IDP can be prescriptive about what's being measured since the state of the application is visible most easily to the platform.

Identifying other critical data

In *Chapter 2*, we defined KPIs to help advance the adoption of the IDP. Those same data points and metrics that drive adoption can be leveraged to help drive innovation and manage technical debt as well. Additionally, when measuring adoption, a natural consequence is that you've measured any failures around adoption. Those failures are a key data point that can help your team understand how developers are relating to the platform.

In *Chapter 6*, how to integrate the platform into the company's way of working was highlighted and we gave the example of a failure to adopt Tekton in favor of GitHub Workflows. This is an example of managing technical debt, as well as an underutilized component that may not be providing sufficient **return on investment** (**ROI**) to justify the ongoing cost of maintenance. Usage metrics such as the number of logins or – in the case of Tekton – jobs scheduled per day, week, and month constitute important metrics. If the platform or a component of the platform is failing to see an increase in utilization over time, especially compared to the platform holistically, then it's time to evaluate that component.

Is the component providing ROI, is it reducing cognitive load, is it meeting SLO targets, and are developers happy with it? If the platform or a component is failing along these lines, then it has more marks on the technical debt side of the ledger compared to the returns.

Data retention is technical debt

While observability data is incredibly useful to decisions for the future of the platform, it's important to keep only as much data as you need. As the amount of stored data increases, you're likely to find that keeping it has diminishing returns and upkeep is a large effort. The more data storage a system requires, the more critical the design of that data storage becomes as it can start to negatively impact system performance as the amount of stored data grows. Additionally, since data is the most critical asset and the most at risk in the event of a security incident, keeping only what is strictly necessary helps manage the surface level of risk and effort involved in measuring the impact of any data leak event. While we authors can't guarantee how much or how little data you need to keep application performance metrics while carefully deciding to keep the data and following the best practices, what we've laid out here and in *Chapter 7* should help ensure the data that is kept is manageable. Something to keep in mind is that the platform, the users, and the technology will all be very different year over year, so data that's retained for longer periods becomes less and less applicable as the platform continues to modernize with the industry. It's best to ensure that any data that's kept has a compelling reason to be there; otherwise, it's just another tick on the debt side of the ledger.

Now that we know more about how we can use data to identify and manage technical debt, let's apply it more practically to our platform.

Maintaining and reworking technical debt

Every aspect of creating an IDP up to now represents a technical debt. Idealists would tell you that if you've made smart decisions, there is no technical debt, but frankly, that's not accurate. A once-thriving open source project can be shuttered with no notice or an unexpected CVE can appear. There will always be debt, regardless of best efforts. What's critical for organizational health is the ability to manage the fallout from such events and maintain a reasonable pace for innovation.

Own your technical debt

Owning your architecture means owning the technical debt that it incurs. While this sounds like an abstract concept, it's very easy to realize the breadth of impact when we start assessing the components of our architecture. Remember this important note from *Chapter 2*?

> **Important note**
> The environment you choose for your platform automatically prescribes some parts of your platform, whether you like it or not! Increasing the number of cloud and infrastructure providers exponentially increases the challenges for your platform design.

The infrastructure required to support the IDP can be one of the most sprawling items on the technical ledger. Ownership means that the debt and the surface area of risks associated with it are well-defined and understood. Sometimes, that risk may mean security or compliance-related risks, but usually, in the context of technical debt, we mean risk to the team's ability to maintain and deliver a product without burning out. It's on the platform architect to embody that ownership, but they can't be a superhero here. Ownership of technical debt, including reacting to unexpected situations, is a team effort.

Technical debt and team ceremonies

The most straightforward way to engage the entire team in the ownership of technical debt is to bake it into the ceremonies the team participates in. The final look of this should be designed by the team. In an agile working environment, this can include a dedicated review of technical debt during the planning phase of each sprint or project, and it can also look like dedicated time per sprint spent on reducing technical debt, such as toil reductions or tackling identified opportunities for improvement.

If your platform team is working in a Kanban fashion, ensuring the backlog is stack-ranked so that technical debt items get picked up among the rest of the work can also be a good strategy.

> **Important note**
>
> There's a very real and very serious risk of these items being constantly de-prioritized in favor of something else. However, this has a snowball effect: the level of effort required to address technical debt will become greater the longer work is deferred.

It's important to develop immutable ways of keeping technical debt under control. Many organizations have different strategies for solving unexpected problems. What's important is to master the sociotechnical aspects of your organization to ensure the scales do not tip out of balance.

Platform composability

The composable nature of the platform lends itself well to being reworked. The components of the IDP should be fairly standalone with few – if any – interdependencies. Thus, the team should feel empowered to decide to leverage this aspect of the platform and deprecate and replace components where appropriate.

The team would need to take this decision carefully as replacing an existing component is a little more complicated than just adding a new one. In addition to answering the questions for evaluating new technical debt, another set of questions must be answered for retiring the old.

Here are some deprecation criteria to consider:

- Do we permanently lose the functionality we need or care about?
- How will this impact the users?
- What would the transition plan be?
- What is the timeline?
- Are we replacing it or just removing it?

- How long would the original and replacement overlap?
- Is there a cost difference? If so, what?
- What are the new maintenance requirements?

Dependencies

Managing dependencies is one of the key areas of managing technical debt. While taking on several solutions that the team doesn't control might initially feel like an anti-pattern for some, the reality is it's typically easier to manage the state of your dependencies than it is to try to write a fully functional bespoke IDP in a reasonable amount of time.

If we look at our dependency matrix from *Chapter 2*, we can apply it to discussions around maintaining the platform and its debt:

Legend: u = unidirectional, b = bi-directional, e = enhancing, c = conflicting

Y depend on X	Tool A	Tool B	Tool C	CNI 1	CNI 2	CNI 3
Tool A					b	
Tool B	u					u
Tool C		b				
CNI 1			e			
CNI 2	b			e		c
CNI 3		u			c	

Figure 9.2: Dependency matrix

Your platform may have a few different types of dependencies. Here are some obvious examples:

- Open source components
- Networking and network appliances
- Databases
- Paid components
- Internal or bespoke services

In a perfect world, an IDP would always be perfectly uniform, regardless of where it was deployed. However, in multi-cloud or multi-architecture deployments, the infrastructure dependencies are likely to be similar but different. As such, you may need to build and maintain a couple of different versions of this matrix to be accurate to the environment. Whatever you decide to do, you will need to ensure these documents are kept up to date to prevent mistakes or security-related issues in the future.

Security

Often referred to as a moving target, security best practices evolve and change with the industry. When those changes occur, your software has fallen behind, and needing to catch up becomes a tick on your technical debt ledger. For example, if we look at workload identities on the major cloud providers, we see a somewhat real-time example of the industry transforming and a new wave of technical debt being identified.

Before, the use of secrets by applications was the norm, and the risks with those secrets were accepted. While mitigating those risks constituted technical debt, there was no better alternative at the time. However, the invention of workload identities removed the need for applications to use secrets to interact with the cloud providers. Instead, they have a role-based identity that they can use to get short-lived tokens, similar to what a human user would have.

Now that this technology has hit all three big cloud providers, there are ongoing efforts across the industry to modernize applications and leverage workload identities instead of secrets. For some applications, this will be possible with a refactor, but for others, this may mean a new component will be added or a full-scale rewrite is required.

Not all technical debt has equal weight

The previous subheading of security is probably the most obvious case of this, but not all technical debt is of equal weight. Anything that compromises the integrity of the system, such as security and compliance, will be very heavily weighted, though other items that may come with a high time cost may find themselves also at the top of the list.

For example, while the release of a new application may be done by hand the first time or two, that quickly gets automated away. From there, automation, which is less timely to kick off and less difficult to maintain, takes over. This is a quick and obvious reduction of the effort required to manage the software, but automation still needs maintenance and that would be considered technical debt. However, since it's unlikely to change often, that debt is less weighty than the release process would have been if left as a manual process.

This is why it's important to regularly review and understand where your team's time is spent when they're not innovating, or what gaps have been identified in the platform's offering.

Now that we've discussed how to get a handle on your technical debt, let's expand on this to understand when that debt should be refactored or rewritten.

Rewriting versus refactoring data – a practical guide

When technical debt snowballs, the team will spend more and more time addressing that debt or the fallout from it, and less time innovating. This looks functionally like higher degrees of operational toil, longer ramp-up and onboarding periods for newer team members and users, and team burnout. It's important to stay on top of technical debt to ensure developer happiness and innovation are both front and center for the platform team.

Sometimes, this snowball can be that things have stopped behaving as expected or took more time than expected. Alternatively, it could be that what was once seen as an acceptable amount of work to run the platform prevents the platform and the team from scaling as the user base grows. Regardless of which exactly it is, as those technical debt items begin to exceed the predefined expectations for normal, the team needs to look at what it would take to return to normal or to find a new good state.

The options for this may be to refactor some of the platforms. A refactor could allow for a fairly quick change for long-term gain. For example, when Golang changed how it did version addressing for dependencies, many teams had to go in and adjust their imports to match the new, better, method. Here are some other reasons to refactor:

- Improve efficiency
- Improve security
- Add or change an interface

However, when there's no way to take the current code base or components list and get to a desired state with it, a rewrite may be considered. We've already established that at its baseline, the system should be composable, meaning few – if any – serverless functions or scripts acting as helpers. One reason to approach a rewrite of a service would be if the continued use of the service would result in the need to add that type of helper workload. **Envoy Proxy** is an example of such a helper workload. Their documentation describes it as "*a high-performance C++ distributed proxy designed for single services and applications, as well as a communication bus and "universal data plane" designed for large microservice "service mesh" architectures*" that "*runs alongside every application and abstracts the network by providing common features in a platform-agnostic manner.*" (Source: `https://www.envoyproxy. io/`). However, for this example, as powerful as Envoy Proxy is, if the only reason to implement it is to support a singular service that is written in another language to the rest of the services, then it may be more worthwhile to rewrite the outlier compared to adding the Envoy Proxy layer.

While adding Envoy Proxy would be an elegant and ready-to-go solution, it's a fairly complex layer to add to the Kubernetes platform, and doing so for only one service would need to be justified compared to that service being rewritten.

Determining whether a rewrite is necessary

Imagine you're on a platform team and you're looking at a component and trying to decide on its future. What would lead you to determine it was time to say goodbye to the component in its current state? One reason could be licensing changes. If the tool you were using used to be open source but has now changed license types, then it might be time to replace it. After all, the open source version may not see any security-related updates, and your team may not have the skills needed to maintain the tool as-is.

The decision tree for rewriting a component should look nearly identical to the decision tree for build versus buy. When you're at the point where a rewrite feels required, the build versus buy conversation needs to be navigated once again. Where an open source solution or an affordable solution for sale exists, your rewrite could be reduced to implementing a new integration. But if that's not the case, then a bespoke service may need to be created.

However, if you're looking to expand functionality or achieve some performance gains, this is very unlikely and an entire rewrite is necessary. Unless the component is super old or hasn't been maintained at all since being written, you're likely going to be better off refactoring your code base.

Examining the external influences on refactoring with an example

Sometimes, the need to refactor doesn't come from within but is a result of a change in a tool or technology within your ecosystem. Unfortunately, the decisions that are made by the maintainers of these dependencies directly influence your team's technical debt and can result in a need to refactor the implementation of that tool.

A good example of a time to refactor would be when transitioning from using Helm (`https://helm.sh/`) version 2 to Helm version 3 for Kubernetes package management. Helm is a package manager for Kubernetes deployments. It's an open source tool that uses a packaging format known as **charts** to orchestrate the process of installing and upgrading applications on a cluster. A chart is not one file, but rather a collection of files that define the Kubernetes resources required to run an application. For example, Argo CD can be installed on a Kubernetes cluster with Helm charts.

Helm 2 was a popular version of Helm until Helm 3 was released in November 2019, but it had some downsides in how it handled **customer resource definitions** (**CRDs**). While Helm today still doesn't support upgrading or deleting CRDs, the new method helps Helm to work better for users leveraging CRDs.

So, what was the problem that Helm 3 solved? First, let's look at what a CRD is. This was explained in greater detail in *Chapter 4*, but if you've forgotten, a CRD extends the Kubernetes API. Functionally, what this means is that your Kubernetes cluster now has a concept of your custom resource. In Helm 2, the handling for CRDs was done via the `crd-install` hook method. The hook mechanism in Helm allows a chart to reach out to another dependency and leverage it before the chart is run. While this should have worked in theory, this still resulted in situations where Helm didn't know about a CRD, so the API extension wasn't understood by Helm to validate a chart. If a new CRD was added after an installation, upgrades would fail because the `crd-install` hook method wouldn't be run. This meant previous versions of the charts would have to be completely blown away and replaced if a CRD changed. This was quite disruptive for users of Helm and many teams had to implement workarounds for this in their CI/CD pipelines.

Helm 3 changed this user experience by deprecating the previous `crd-install` hook method and removing the underlying functionality that supported it. Now, a CRD directory is needed. The directory path is nested within the chart, and the use of this directory, called `crds/`, allows Helm to pause while the CRDs are added to a cluster before continuing with the chart's execution. Helm upgrades do allow for CRD installation, whereas they didn't before. Essentially, Helm now knows about new API functionality when it does its chart validation, ensuring a smoother operations experience for users.

While the underlying Helm changes were pretty significant, for users of Helm, this equated to refactoring existing charts and upgrading the version of Helm in use. However, this also allowed many users to remove whatever workarounds they had implemented and instead allowed them to leverage Helm's built-in upgrade functionality. Depending on the size and complexity of the charts being used, these refactors could have been a somewhat significant effort, but nowhere near as much as migrating to a different way of handling software installations and upgrades altogether.

To understand the rough scope of the refactoring changes, let's look at today's Argo CD Helm Charts using Helm 3 (`https://github.com/argoproj/argo-helm/tree/main/charts`). Within the repository, there's a collection of charts that are each installed as part of the overall Argo CD project. With Helm, directory structure is very important; without the correct structure, a chart can't function.

If we look at one of the charts, we'll see that the new CRD directory is present:

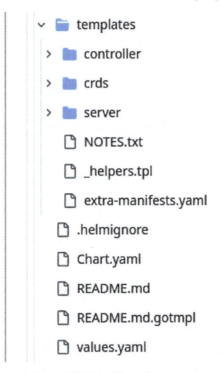

Figure 9.3: Argo CD Helm Chart directory structure

If we take a look at the `values.yaml` file within one of these charts, we'll see that the CRD usage is very straightforward:

```
1    images:
2      # -- Common tag for Argo Workflows images. Defaults to `.Chart.AppVersion`.
3      tag: ""
4      # -- imagePullPolicy to apply to all containers
5      pullPolicy: Always
6      # -- Secrets with credentials to pull images from a private registry
7      pullSecrets: []
8      # - name: argo-pull-secret
9
10     ## Custom resource configuration
11   crds:
12     # -- Install and upgrade CRDs
13     install: true
14     # -- Keep CRDs on chart uninstall
15     keep: true
16     # -- Annotations to be added to all CRDs
17     annotations: {}
```

Figure 9.4: CRD configuration in a Helm Chart values.yaml file

The real bulk of the work to leverage CRDs from a development perspective is within the CRD definitions themselves. Here, the CRD annotations are defined and any other logic that's required, such as the resource policy, is stated:

```
1    {{- if .Values.crds.install }}
2    apiVersion: apiextensions.k8s.io/v1
3    kind: CustomResourceDefinition
4    metadata:
5      name: workflowtemplates.argoproj.io
6      annotations:
7        {{- if .Values.crds.keep }}
8        "helm.sh/resource-policy": keep
9        {{- end }}
10       {{- with .Values.crds.annotations }}
11         {{- toYaml . | nindent 4 }}
12       {{- end }}
13   spec:
14     group: argoproj.io
15     names:
16       kind: WorkflowTemplate
17       listKind: WorkflowTemplateList
18       plural: workflowtemplates
19       shortNames:
20       - wftmpl
21       singular: workflowtemplate
22     scope: Namespaced
23     versions:
24     - name: v1alpha1
25       schema:
26         openAPIV3Schema:
27           properties:
28             apiVersion:
29               type: string
30             kind:
31               type: string
32             metadata:
33               type: object
34             spec:
35               type: object
36               x-kubernetes-map-type: atomic
37               x-kubernetes-preserve-unknown-fields: true
38           required:
39           - metadata
40           - spec
41           type: object
42       served: true
43       storage: true
44   {{- end }}
```

Figure 9.5: A CRD from the Argo CD project

After seeing how CRDs are used with Helm charts today, a natural question would be, *"What did it look like before?"* The answer is *"It depends."* Before the `crd-intall` hook was added to Helm, a common way of working around this issue was to actually break out CRDs into separate charts and then use a bash script to ensure charts were installed in a specific order. The Istio project is an example of where this was done. This worked but created more overhead to maintain since the scripts for ensuring proper installations were yet another thing for Helm Chart maintainers to keep up to date.

After Helm added the install hook, the ability to deprecate such scripts became possible. While keeping the same CRD files, a user could add the crd-install hook annotation. Then, Helm's underlying component tiller would ensure that the CRD creation was triggered before the rest of the chart was actioned.

The Helm 2 docs explain this further (https://v2.helm.sh/docs/charts_hooks/#hooks). The essential thing to know is that a YAML file containing the crd hook would look like this initially:

```yaml
apiVersion: apiextensions.k8s.io/v1beta1
kind: CustomResourceDefinition
metadata:
  name: crontabs.stable.example.com

spec:
  group: stable.example.com
  version: v1
  scope: Namespaced
  names:
    plural: crontabs
    singular: crontab
    kind: CronTab
    shortNames:
    - ct
```

After, it would look like this:

```yaml
apiVersion: apiextensions.k8s.io/v1beta1
kind: CustomResourceDefinition
metadata:
  name: crontabs.stable.example.com
  annotations:
    "helm.sh/hook": crd-install
spec:
  group: stable.example.com
  version: v1
  scope: Namespaced
  names:
    plural: crontabs
    singular: crontab
    kind: CronTab
    shortNames:
    - ct
```

In itself, this is a very small change. The Helm version in use needed to be updated as well, but overall, this was a very small amount of work that allowed many custom bash scripts to be taken off the plate of people maintaining Helm Charts for their applications.

Sometimes, the process of refactoring is pretty small. It can be as simple as identifying a few areas of inefficiencies or a place where the code could encounter race conditions and making changes to the code so that those situations are improved. Other times, the refactoring could be more sweeping, such as replacing a library or updating the language version to a version with some breaking changes. Whatever the reason for refactoring, it's typically less effort with a higher return on investment than a rewrite.

Examining a famous rewrite

If it's still unclear when a rewrite may be the appropriate choice, we can look to open source software projects for some famous rewrite scenarios. Within the Linux kernel, as well as container technology itself and therefore Kubernetes, we can find a perfect example in cgroups. Cgroups, also known as control groups, are the technology that exists within the Linux kernel to manage computational processes. Cgroups manage resource allocation and generally control processes. Recently, however, in response to community calls for performance improvements, there was a rewrite, and cgroupsv2 was born.

This example is particularly interesting because of the pile-on effect. While cgroupsv2 is generally considered superior to v1, and most applications will see performance gains, legacy applications may suffer. In Kubernetes, where one cgroups version or the other must be specified, the issue with legacy applications is apparent. In particular, older Java and Node.js applications will have compatibility issues with the memory querying functions of cgroupsv2 and are likely to experience **out-of-memory (OOM)** errors crashing the workload. For companies using more dated applications, an upgrade of the stack is indicated. This could mean either a rewrite or a refactor, depending on how far behind the legacy code is and what's needed to bring it up to a compatible version with the new cgroupsv2.

Those with more knowledge of Kubernetes might note that as of current version 1.30 of K8s, Kubernetes still contains the capabilities of using cgroups v1. This is true, and any organization that doesn't wish to modernize can leverage that capability. However, counting on the continuation of cgroupsv1 over the rewritten version would be remiss as it's more likely that v1 will be deprecated instead of two versions being maintained. If a modernization wasn't desired, then instead of a new piece of technical debt, the cluster configurations that specify cgroupsv1 would be added to the team ledger. All new clusters must be created using the cgroupsv1 specifications for as long as it's supported or for as long as the legacy applications aren't modernized. If there's any automation in creating the clusters, then that would need to be refactored and added to the additional configuration. Additionally, while more modern applications will be able to run with cgroupsv1, they may not be as performant as they might have been with cgroupsv2. Regardless of whether the decision was to refactor cluster creation or to modernize workloads that were running on older language versions, the effects of this rewrite rippled across the industry.

A final consequence of the rewrite and subsequent adoption of cgroupsv2 as the default is that some new innate Kubernetes features are only compatible with the newer version. In other words, you limit your capabilities by sticking to cgroupsv1 and miss out on leveraging some of the Kubernetes enhancements for workload management.

Transitioning after rewrite

You've done the rewrite (or maybe replacement) and it's time to switch things over. Unfortunately, these types of transitions are rarely – if ever – clean and typically require precise planning and execution. The messier the process, the longer the original component must stick around, and the longer the team must balance the operational costs associated with both. To help foster a healthier transition, define a transition plan early and test your transition plan, if possible, to ensure nothing was missed. Additionally, there's another benefit to writing out and testing a transition plan. The completed plan and tests become a record of the events and a blueprint for future migrations that may need to occur. While we always hope a rewrite is something that only happens once, the reality is that in any business that survives for long periods, multiple rewrites will occur. By documenting this process, you help curb the technical debt of the rewrite. This will help future engineers untangle the history of the system and save them time generating transition plans in the future.

Architectural decision records – document for the Afterworld

An **architectural decision record** helps future owners of the platform to understand not just what was done, but why it was done. Maintaining accurate architectural diagrams and documentation is critical to understanding any software system, but the reasoning behind those decisions and final states help future leaders know what to do with the platform in the future.

Why document software architecture?

Software architecture should be documented for several reasons. First, it can help onboard new team members and users to the project. A good architecture document helps these people understand how the system is used and how data flows through it.

Second, it can aid with security and compliance as all security and compliance audits require thorough software architecture documents and diagrams.

Finally, it gives you a starting point from which to grow. By documenting the current state, you provide a reference document to plan against. How could a future architect figure out where the platform should go without a good understanding of where it is?

What does good technical documentation look like?

A good architecture document will have a mixture of images and descriptions. Depending on the granularity of the document, it should show how the users interact with the system, the dependencies of the system, how data moves, how data is stored, where data is changed (if at all), and what the expected outcomes of interactions are. In other words, the documentation should capture what the system does and how it does it.

An additional function of the documentation is to record not just the what and how but also the why. Possibly more important than the what and how, the why behind each decision helps guide future decision-making processes. For example, the architect of tomorrow will need to know key information about the decisions made:

- Was there a limiting factor that caused a less-than-optimal solution?
- What bar would a replacement solution need to reach to be considered?
- What were the known risks and mitigations at the time?
- What were the goals and non-goals and why?
- What impact does each component have on another?

By ensuring you record these aspects of a decision, you're building a legacy, not legacy software.

However, as critical as architectural decision records are, they are not the only important document for your platform. Other types of documentation mentioned in the chapter and book should be compiled and maintained so that you stay on top of the technical debt the system incurs. While managing the documentation's organization and keeping the documentation updated can feel like toil, and is indeed debt, the time savings gained by doing so pays dividends over time. Searchable, well-organized documentation is the key to a system that stands the test of time.

Our fictitious company – a final look

ACME Financial One is a company that's designed for our purposes – to feel like many real companies modernizing their technical stacks. In our example, they're dealing with both legacy systems and new or greenfield systems. As with any real company, this would result in an initial duplication of technical debt. To support engineers during this transition period where two systems exist, a well-designed platform will provide the flexibility required for smooth transitions between the two environments.

ACME Financial One would need to support this with the thinnest viable platform that scales and grows to meet users where they are. Clear documentation of the platform and the processes for leveraging the platform will support the developers in their transition and their adoption of the new environment.

Summary

If there's one lesson to take away from this chapter on technical debt, it's that asking the correct questions at the correct time can help your team build a platform that's designed to last. The platform may never be *finished*. There will always be improvements to be found and gains to be made. However, a polished process for managing the platform and its corresponding documentation will allow the team to sustain it while maintaining the team's health.

As you're tackling technical debt, remember the example indicators and questions we discussed in the introduction to this chapter. Spotting technical debt and knowing the corresponding costs will allow you to better communicate and prioritize addressing those items with those teams that need to address them. While addressing technical debt is time-consuming and costly on its own, it will help you set the foundation for a platform that can last and is ready for future changes.

In the next and final chapter, we will discuss changes and how your platform can provide these changes as a stable foundation for your IT organization. You will learn about the idea of sustainable and lightweight architectures and how you can enable your users to follow those approaches. We will challenge the perspective of the golden path as a user-oriented term and question whether we, as platform engineers and architects, don't also require a golden path. Finally, we'll close the book with some final thoughts about the next trending topics.

10
Crafting Platform Products for the Future

The difference between a project and a product is its lifetime. Projects have an end. If they pass their deadline, organizations reallocate resources, money, and time, leading to the degradation of a system. A platform as a product must be built to adjust, evolve, and grow with changing requirements over time. Platforms are also more than just a technical solution. They require a team, a mindset, and a whole organism to bring pure value to a company.

In the final chapter, we want to find some encouraging words that encapsulate the book's learnings and insights. We will discuss the continuity of changes and how your platform can provide this continuity as a stable foundation for your IT organization. We will then look into the principles of sustainable and lightweight architectures, as well as how you can encourage your users to follow those approaches. As a platform engineer, it is up to you to unlock these capabilities and add value.

We will close the topic of this book by challenging the perspective of the *golden path* as a user-oriented term and questioning whether we, as platform engineers and architects, do not also require a golden path.

After some final thoughts in this chapter about the next trending topics, we will let you go to keep or start working on your platform. You might have realized that we often wrote "your platform" – a platform as a product starts with your mindset. You are responsible for its quality and the desire it brings to a user. You define the success of it, and it is nothing that should be discarded afterward. That's why it is *your* platform.

In our final chapter, you will learn about:

- Continuous changes – learning to age and adapt
- Considering sustainable and lightweight architectures and support
- The golden path for changes
- A glimpse into the future

Continuous changes – learning to age and adapt

The major challenge of ongoing change is that as a platform team, you need to adopt and embody a forward-thinking mindset yourself, while also guiding and inspiring your users to adopt the same approach. If you can't provide this, in the best-case scenario, your users will request and force your solution and team to adapt. In the worst-case scenario, they will abandon your platform and will look for other solutions. So, it is down to you to provide a solid but flexible foundation for change and growth.

The imperative of change

We previously discussed how fast technology evolves, as well as how major technology milestone cycles happen every 10–15 years. As Kubernetes has just turned 10, we can expect a revolution in the next few years. However, smaller changes also create the necessity for platforms to adapt to new requirements and innovations. Platform architects have to find a way to balance the stability of a platform and the innovation given by the market. This tension between those two elements can be overwhelming. It will question the concepts of your platform, the decisions of your IT organization, and the design choices of your users.

The MIT Sloan School of Management has defined this underlying structure of continuous change. They found out that what often looks like chaotic changes is just a continuation of the four cycles shown in the following figure. It took them 15 years of research to understand that these changes do not happen linearly but start over and over again. Where do we start in this cycle? It depends because we are already in it, and actually, every technology, methodology, or framework has its own cycle, slightly influencing other cycles. Let's take platform engineering as a concept. Where it really started is difficult to say; did it evolve from DevOps? Did the term come up with tools such as Backstage? Was someone thinking philosophically about the IT world and thought that it was a good idea? Most likely, it was a little bit of everything that started the self-propelling force of platform engineering. It led to new tools and architectural perspectives. **DevRel** (short for **Developer Relations**) professionals discussed it with others, demonstrating how it works, which eventually led to us writing a book about it.

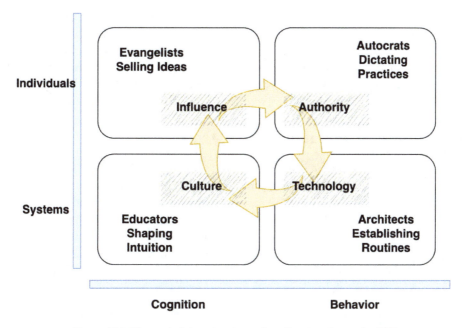

Figure 10.1: The underlying structure of continuous change by MIT

If you see such self-propelling behavior that is in balance across all four domains, then it will likely be adopted by *everyone*, and you can take a step as a first mover.

Fostering a culture of change

What do we take from the imperative of change? We cannot avoid change; if you do so, you avoid your own and your platform's future. As a platform team, you are in an ideal position to foster a culture of change. Yes, yet another topic to put on your plate, which has nothing to do with engineering, but that makes the role of platform engineers so special. You are in an ideal place because the platform and your work are the connections between your IT landscape, the development teams, and the business.

Taking this responsibility and becoming a thought leader isn't an easy task. To add the role of platform engineers who set a vision for the future and lead teams through change requires strategies to build a team culture that embraces change, including communication, incentives, and leadership support. The right cultural approach depends on your team and the individuals in your organization, something that barely can be covered in a book. What we can recommend is to create environments that invite users for experimentation, such as sandbox environments or pilot projects. But you have to do more; you need to create such environments for your own team too, and you need to encourage your platform engineers to test new technologies. Your results must be documented and available for the whole team so that you avoid duplicating work, as well as so that you can compare the maturity of tested technologies over time.

Therefore, you have to implement a routine to celebrate success and failure in the same way. In practical science, failure is an important part of research. Companies such as SpaceX blow up rocket after rocket, but they still count it as a success because, every time, they come closer to their target. Maybe don't blow up your data center, but at least allow a safe environment where you can invest money for failure. Organizations waste money all day long for seemingly nonsense activities. If you can do something that makes your team and your technological readiness grow, it is way better to do so.

Iterative evolution and an evolving strategy

With a safe environment for testing new technologies, you will be able to identify which ones to introduce to your platform and maybe even replace an old one. Finding the balance between stability and innovation can vary from case to case. We have seen platforms that always release, in a big bang, a completely new environment and become very professional in automatically migrating the workload to the new platform. However, many decide to take the approach of introducing new technologies and features step by step. Utilizing feature toggles and giving access in a canary deployment style to a user allow you to slowly but safely introduce changes. With good observability and feedback loops, you will also get the chance to raise the quality of the new features and let the user be part of the evolution.

Evolution doesn't need to happen with big milestones; it is also important to continuously do the following:

- Refactor code and improve the quality of it
- Take feature requests into consideration
- Reflect on technical debts and see whether they have since become solvable
- Update third party integrations frequently

In this context, it is also important to think about how to deprecate implementations. Some tools might have experienced a lot of hype but died soon after being implemented, due to a missing sustainable contribution and development. Others might be replaced over time with new major versions. Setting up a depreciation strategy is, therefore, also a part of the evolution of your platform. It requires clear communication of the impact and timelines to the user. With deprecation, you will also identify how modular and loosely coupled the components of your platform really are and whether users have built up some kind of dependencies.

Adopting new technologies is a journey and a mindset. As platform engineers, you are in the right place to do so. Establish a continuous learning environment, frequent feedback loops, and post-mortem reviews to learn quickly and speed up successful evolution.

In the next section, we will shortly highlight a recent trend in sustainable development. We will focus on environmental sustainability, which has become a strong driver within the last years, and the sudden growth of **Generative Artificial Intelligence** (**GenAI**) has put the topic under the spotlight ever more. So, here are a few things we can do as platform engineers to deliver our part for a sustainable future.

Considering sustainable and lightweight architectures and approaches

IT has become one of the global top **Carbon Dioxide (CO2)** originators, alongside the construction and aviation industry. The best studies on it are, unfortunately, several years old, but we can summarize the 2018 findings:

- The global IT consumed around 2% of electricity

- This caused between 1.5%–2.4% of the global CO2e (**e** stands for **equivalent** –not every gas originated is CO2)

- This range is comparable with the CO2e share of Indonesia (1.48%), Canada (1.89%) or Germany (2.17%)

- Predictions in 2018 assumed that these numbers would double by 2025; some less trustworthy sources even predict 12% increase in CO2e by 2025

We don't have to shut down everything and start living like cavemen again, but we should at least become aware of the impact of our *infinite* computing resources on the environment. Computing resources, of any kind, causes the output of CO2e during their production, their usage via the electricity they need, and when retired and destroyed.

The best things we can do, therefore, are as follows:

- **Reduce**: Reduction is all about being sustainable about what we consume. In our daily lives, we don't have to take plastic bags, but what do you do with digital resources? Well, it's simple – do we really need an ultra-highly available web server distributed over five continents with hourly cross-location backups?

- **Reuse**: In IT, reuse has two aspects. If a server gets retired, it might make sense not to destroy it but to sell it to a provider who specializes in running legacy servers. Second, we should keep using the server as long as possible. This is also a form of reusing it. Kubernetes, as an orchestrator, makes this extremely simple and enables us with well-proofed balancing if there is a failure.

- **Repair**: Most of us do not have the capability to repair servers. Often, individual components such as memory modules or network devices fail.

- **Rightsize**: Adjust the requested computing capacities, on-premise and in the cloud, to fit your workload. The era of 80% headspace on a server is over. Those who don't use their compute resources at least 90% are practically harming the planet.

- **Reject**: Don't work with providers who are not committed to reaching sustainability targets. It is up to you to add this as a criterion to your vendor selection processes.

- **Replatform**: Be aware if a vendor or platform is not committed or doesn't support sustainable targets.

Now, you might say that the business case for "sustainability" is not very economically interesting. You might even have gotten the feeling that we discussed actions we should take previously. Reducing costs often comes along with optimizing your CO2 footprint. Therefore, optimizing your footprint also positively impacts your costs. Further, more and more regulations are being shaped to achieve a slower-growing global warming. Fulfilling those regulations is better than paying fines.

Enable lightweight architectures

Enabling lightweight architecture should be a primary target for a platform. It not only has benefits in regard to sustainability but also to its robustness and the ability to forgive failure. The architecture of an application becomes lightweight when it doesn't have hard dependencies, utilizes fewer components, is flexible in its deployment (which includes scalability), and uses just the right technologies. A good example here is Apache Kafka, which became a kind of de facto standard in the context of stream processing. However, I have seen more systems using it just to transport logs and events and consolidate data flow. There are dozens of more lightweight and efficient tools out there that provide the same capabilities, just without the scale, the resource demand, and the overhead of complexity. This pattern is very common. We prefer to take a solution that answers our demand and can do so much more, even though we don't need it although others do, so it has to be the right choice.

Whatever we choose, to enable lightweight architectures as a platform team, we have to provide technological options to be able to provide two to three better answers to a demand, rather than the wrong one.

Providing support for the users to adapt

How do we support users now and allow them to optimize their systems going forward?

We give them transparency about their impact, we show them the possible platform capabilities to scale, balance, and run serverless, and we provide alternative lightweight technologies. This can be an integrated part of our **Internal Development Platform** (**IDP**) and observability.

To create transparency of the carbon footprint, open source tools such as Kepler, Scaphandre, as well as commercial solutions provide insights into which system component causes which CO2e output. The following screenshot from Kepler shows a simple Grafana dashboard with the energy consumption per component.

Figure 10.2: A Kepler dashboard with the energy consumption per component (The image is intended as a visual reference; the textual information isn't essential.)

Currently, those tools are still under development and often error-prone and difficult to handle, but they are a starting point. Whether they succeed or not will depend on the ability to educate users. We have to change users' perspectives from building software in one environment to a platform that can do it for them. Software needs to be built on a platform to natively use all its capabilities. That might sound like we want to build and introduce dependencies, and with that, break our concept of modularity. But this is the difficulty of a product; you have to find the right approach to evolve and enable new approaches while keeping backward compatibility. If your new features or the replacement of an old feature brings value, users will adapt to it.

> **Important note**
>
> The best way to adapt new approaches is achieved through openness, which fosters transparency and education.

In the next part, we will wrap up our platform journey with the golden path. Once again, we will take a look from a different perspective at one of the common topics in platform engineering, considering a golden path as a feature for a platform. Ultimately, as platform engineers, we create a golden path to create more golden paths.

The Golden Path for changes

The term *Golden Path* has been used several times throughout our book. One of its first uses was in a **Spotify** blog post in 2020 *[1]*. Golden Paths were created to guide engineers through supported ways of getting things done (e.g., using a particular service or creating a new piece of code that had a certain purpose). Platform engineering is all about providing easy-to-use *Golden Paths* to reduce the effort for individual teams to create, compile, test, deploy, and operate software.

Looking once again at our fictious company Financial One ACME! Earlier in the book we identified several use cases throughout the book that we can provide as *self-service Golden Paths* to engineers that we, as the platform engineering team, support. A key aspect of providing good support for any product is proper feedback channels, which is what we will discuss next.

Providing feedback channels

Supporting our platform means that we make sure that all golden paths provided through it work as expected, all the time. Should something not work as expected, the platform engineering team needs to make sure that we fix whatever is broken as fast as possible, mitigating a negative impact on our end users.

Ideally, we automatically detect whether things don't work as expected and don't have to wait until our users are complaining. In *Chapter 3*, in the *A Platform that is available, resilient, and secure* section, we discussed how important it is to observe our platform and all its components to automatically detect whether things are not working as expected. Like with any other software product, we can and have to use observability to ensure our platform and all its components meet their **Service-Level Objectives (SLOs)**. So, if our observability tells us that something is not available, has slowed down, or is all of a sudden throwing error, then we must act on it to return it to a working state.

Another way to get automated feedback through observability is by observing how many users actively use the platform and which golden paths they follow. For a refresher on this topic, revisit *Success KPIs and optimization* in *Chapter 3*. If we see that, all of a sudden, our users change their usage behavior, then that's an indication that something has changed that we need to address. It could be that some of the golden paths are broken (e.g., developers can't access tools such as Backstage, Git, or ArgoCD), or it could be that some paths are outdated (e.g., a Backstage template is not working anymore because it uses an outdated configuration). It could also be that the requirements, regulations, or preferred technology stacks in an organization have changed and that our platform currently does not provide golden paths to reflect new use cases that accompany a change in requirements.

To have an up-to-date set of Golden Paths, we need to provide a *feedback channel* where our users can actively give us feedback on new requirements they have. It's like requesting new features on any software product. The best way to learn about those new feature requirements is to open a direct channel to our end users who are our internal engineers. This can be done through an internal chat (e.g., have a dedicated Slack channel such as `#goldenpath-suggestions`), you could provide an email distribution list (e.g., `goldenpaths@yourcompany.com`), or you could build a *self-service feedback Golden Path* into the platform itself that users can use to provide details on what they would like the platform to do next.

Feedback is the ultimate driver to continuously improve our platform with new or improved golden paths. For our end users, this means they continuously get new features from the platform to make their day-to-day lives easier, even as their requirements and processes to deliver software may change.

Golden Paths are the features of our platforms

As mentioned in the introduction of this final chapter, a platform is a product, and as such, it is never finished, in contrast to a project, which has a final deadline. A platform evolves, as there will be new use cases that need to be added to it when new golden paths or existing golden paths have to be updated to keep working, due to changes in requirements.

As we are building our platform like a product, we can also think of our golden paths as the features of our product. Those features will come in different shapes or forms, depending on how we decide to build our platform and which types of user interface we provide to our end users. In *Chapter 3* in the *A reference architecture for our platform* section, we discussed the different forms of user interface for a platform – **Command-Line Interface (CLI)**, a REST API, a special config file in a Git repo, or a nice-looking developer portal UI that lists all available use cases. An example of such a user interface that provides self-service access to a list of available Golden Paths (i.e., features) is shown in the following screenshot:

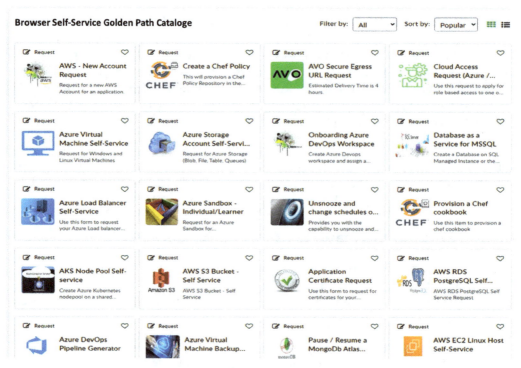

Figure 10.3: A self-service portal with all the available Golden Paths (i.e., features) (The image is intended as a visual reference; the textual information isn't essential.)

The preceding screenshot is from a financial institution (similar to our Financial One ACME) that provides a list of self-service golden paths across a wide variety of use cases, which either the platform engineering team has identified or that their end users have requested over the years. Most of the visible golden paths in the screenshot focus on requesting cloud resources (e.g., EC2 or Azure Virtual Machines), requesting cloud services (e.g., AWS **Relational Database Service** (**RDS**) or Azure Load Balancer), or creating new things based on a template (e.g., creating a Chef policy or an Azure DevOps pipeline).

When clicking on one of those Golden Paths, the user is typically prompted for some input, which will then trigger some automation to fulfill the request. Some Golden Paths that cannot be or are not yet automated might just provide a how-to guide, explaining how to get this specific request fulfilled.

To get a refresher on the impact of those golden paths in real life, go back to *Chapter 8,* which focuses on cost optimization. In the *Only request what you need approach* section, we explained what happens behind the *requesting an Azure Virtual Machine* Golden Path.

The preceding example is already a good overview of how a platform can evolve. But we don't start out with that many Golden Paths (i.e., features). We start small and then need to be ready to scale. But how do we do this?

Start small, expect growth, and don't become a bottleneck

In *Chapter 2*, in the section about **Thinnest Viable Platforms** (**TVPs**), we discussed how we need to start with a small set of use cases that we believe will make an impact. From there, we use feedback to learn more about additional use cases that we use to grow a platform.

Let's look at Financial One ACME. In the first chapters of this book, we discussed several self-service use cases that would improve the lives of developers, DevOps, ITOps, or testers at Financial One ACME. That included use cases such as *self-service access to production logs for troubleshooting*, *provision compliant environment*, *running performance and resilience tests*, and *onboard of a new application*. Once those use cases are implemented and proven to make the life of those engineers easier, it's guaranteed that more *Golden Path Requests* will appear. So, we need to expect growth that comes out of building a successful product.

The problem with a growing set of use cases that needs to be supported is that it requires more effort to test and maintain, leaving less time to add new use cases or core capabilities to a platform. We run the risk of becoming a bottleneck, as we struggle to keep the platform and its features up to date with all the requirements that our users have. One solution to this problem is to provide a Golden Path to build Golden Paths, making it easier to extend the platform – not just by experts but by everyone!

Golden Paths to Build Golden Paths

Throughout the book, we discussed providing golden paths to reduce the cognitive load of engineers to get their work done. The questions now are, as we design our platform, can we also provide golden paths that allow the platform engineering team to become more efficient in extending the platform with new features, capabilities and use cases? Can we build a Golden Path that allows the platform engineering team to create more Golden Paths? And could that Golden Path then also be used by our end users (the engineering teams of our organization) to implement their new requirements without the need for the platform engineering team as a new Golden Path? It almost feels like the 2010 sci-fi movie *Inception*, starring *Leonardo DiCaprio*.

The good news is that we don't need Hollywood to achieve such a thing. In *Chapter 5*, in the *Providing templates as Golden Paths for easier starts* section, we discussed the templating feature in Backstage. Templates in Backstage are a way to define Golden Paths. The templates are typically created and maintained by the platform engineering team, as their creation requires some expert knowledge. The following snippet shows an excerpt of such a template definition:

```
apiVersion: scaffolder.backstage.io/v1beta3
kind: Template
metadata:
  name: self-service-app-onboarding
  title: Onboarding of a new App to Financial One ACME
  description: Golden Path to onboard a new App
spec:
```

```
owner: financialoneacme/platform-engineering
type: service
# Input parameters needed when using this golden path
parameters:
  - title: What programming language
    properties:
      language:
        title: Programming Language
        type: string
        description: Which language do you want to use?
      owner:
        title: Owner
        type: string
        description: Which team owns this new app?
  - title: Choose a target Git location
    ...
# Steps when Golden Path gets executed
steps:
  - id: fetch-base
    name: Fetch Base
    action: fetch:template
    input:
      url: ./template
      values:
        language: ${{ parameters.language }}
        owner: ${{ parameters.owner }}
  - id: publish
    name: Publish
    action: publish:github
  - id: register
    name: Register
    action: catalog:register
# Output to show back to the end user
output:
  links:
    - title: Repository
      url: ${{ steps['publish'].output.remoteUrl }}
```

For a complete example, take a look at the **Backstage template** documentation *[2]*. The preceding example defines a golden path that allows a user to select a programming language, define who the team is that creates this new component, and a target Git repository where the new project for a new app should be created. Let's assume that a user selects **Java** as their preferred programming language. The fetch-base step then fetches a set of files that have been specified in a git repo for use whenever somebody selects Java. The publishing step then publishes the new git repository, based on the selected language. There is a bit more magic happening behind the scenes, but essentially, this is what creating a Golden Path template is all about.

What should be obvious when looking at this example is that creating Golden Paths is actually not that hard. Sure, there are some specifics that we can't cover in this short section of the book, but it shows that tooling such as Backstage already exists that makes it very easy to create or update Golden Path templates. This also allows platform engineering teams to enable other teams to help create and update their own Golden Paths, thereby reducing the chances that the core platform engineering team becomes a bottleneck.

While we use Backstage as an example, there are also other tools out there that provide similar flexibility. You should make yourself familiar with the tools available in the open source space, and also look into commercial offerings. One additional open source project to have a look at is **Kratix** – *the open source platform engineering framework [3]*. Kratix provides a framework for building composable **Internal Development Platforms (IDPs)**. It has an interesting approach, using so-called *Promises*, which are essentially analogous to *Golden Paths*. Kratix encourages all users to create and contribute their own Promises to the platform, thereby enabling every user of the platform to leverage Golden Paths created by anyone in an organization.

Now that we have discussed how Golden Paths are not just the start of your platform but also evolve, as well as how you can provide Golden Paths to build new Golden Paths to ensure the future growth of your platform, it is time to look into some other future trends we need to consider as we craft platform products that are fit for the future!

A glimpse into the future

The next lines might be the most dangerous ones written in this book. They could be complete nonsense, just a dream, or maybe the future of our day-to-day work. Building a platform means keeping it up to date as things around us change. This includes following the market closely, observing what is happening, and evaluating whether one of the trends is likely to last. Most important of course is answering the question of whether those new trends will be beneficial for your platform's users and organization.

The replacement of hypervisors

The first topic to philosophize about is the potential replacement of hypervisors with Kubernetes. We touched on this topic earlier, especially looking into the history and reasoning behind still using hypervisors. What needs to happen to make this a considerable approach calming down the **Chief Information Security Officers (CISO)** out there? What needs to be solved are things such as the following:

- Container isolation and encryption

- Network isolation and encryption

- Storage isolation and encryption

In short, we have to isolate and encrypt everything from each other, while keeping the container orchestrator fully flexible to provide all required capabilities for the platform. But perhaps we are thinking about this in a too-classic, old-school way. Obviously, we can harden containers to a point where they become an entirely sealed black box. For some applications, we can consider WebAssembly as a compile target that is very secure. And if that is not enough, we can add confidential computing to the whole story or just wait for the open source community to find better solutions. It's as if no one ever breaks out of the context of a VM *[4]*. Isolating networks is also no silver bullet. Technologies such as **Extended Berkeley Packet Filter (eBPF)** are used to "*dynamically program the kernel for efficient networking, observability, tracing, and security.*" eBPF enhances secure network communication by allowing the execution of custom, sandboxed programs in the kernel space without modifying the kernel code. This capability enables fine-grained control over network packet processing, filtering, and monitoring, which helps enforce security policies and detect malicious activities in real time. The last issue would be the storage, but in this space, we build on software-defined storage. Overall, replacing VMs with Kubernetes is doable. For sure, we have skipped a couple of details, but in a cloud-native space, there is a solution for every problem.

AI for platforms

No other topic at the time of writing is so omnipresent as AI, or to be precise, GenAI. Every conference, every media source, and every professional magazine is covered by GenAI. Also, the open source community wasn't inactive and developed K8sGPT as a response to it. K8sGPT is a tool to scan your Kubernetes clusters, diagnosing and triaging issues in simple English. This is the present, but what we will see in the future is interesting. The actual problems of right-sizing, adjusting scale, configuring load balancing, and any other operational task are perfect issues for AI to handle. We have standardized schemas and communication patterns. Methods of finding the best-suited node or the correct configuration of a Pod in terms of its resource consumption should be therefore simple. It is surprising that, currently, we don't see any market-taking solution. But maybe the problem is that it is too simple to use AI. Conversely, we see many different influences on IT as demographics change, and there are many skill gaps and cost pressures at companies and in the market, which might accelerate the AI space for operation.

OCI Registry as Storage and RegistryOps

We are constantly changing how we do our operations. Since the emergence of DevOps, we ops-ify everything – ML, security, cost management, and so on. Another addition to this is RegistryOps. The idea here is to store all artifacts needed for deployment in the same location, rather than having them in various other storage solutions. For this, the Open Container Initiative has defined multiple standards for registries to comply with, such as the distributions specification, runtime specifica, and the image specification. Back in *Chapter 5, The Importance of Container and Artifact Registries as Entry Points*, we discussed this in depth. This makes the registry a single point of truth and failure. The **OCI Registry as Storage (ORAS)** project builds on those standards. It allows the use of container images, building relations between artifacts, and attaching, for example, a **Software Bill of Material**

(**SBOM**) to a container. Of course, the SBOM is stored in the same registry. GitOps tools then use the registry as a desired state store and handle the deployment toward the cluster. As most registries are compliant with that approach, ORAS is simple to scale, provides a very fine-grained permission specification, and improves the supply chain security.

There are more and more supply chain attacks happening, and the supply chain becomes a relevant attack surface. With a registry that lies between CI and CD, you establish a desired breakpoint for those chains. That slows down the process of scanning artifacts for vulnerabilities and ensures that they are signed by the right people. Ultimately, RegistryOps is still GitOps, but it uses a registry and not Git.

Containerized pipelines as code

A fairly new concept or idea is to define a CI/CD pipeline in a container so that you can run this locally, as well as in a tool such as GitHub Actions or Jenkins. By doing that, you can harmonize the automated steps and ensure that it can run anywhere. The idea was introduced by the open source tool Dagger, initiated by the former CTO and founder of Docker Solomon Hykes. In addition to that, you practically code the pipeline, which allows it to be independent from the CI/CD systems' own logic. The following code is just one function from a Dagger file and executes the build of an application.

```
func (m *HelloDagger) Build(source *dagger.Directory) *dagger.
Container {
    build := m.BuildEnv(source).
        WithExec([]string{"npm", "run", "build"}).
        Directory("./dist")
    return dag.Container().From("nginx:1.25-alpine").
        WithDirectory("/usr/share/nginx/html", build).
        WithExposedPort(80)
}
```

This approach can be a game changer for companies that have a fragmented landscape of tools and processes, and who tend to migrate their landscape to something different every other year.

However, some critics also suggest that it might not be necessary to execute CI/CD components locally, as a major trend is to push everything remotely to optimize resource utilization. Platforms such as GitHub Actions provide reusable steps that just need to be filled with values. Let's see if we will dagger in future pipelines.

Platforms – a better future with them?

With platforms, a product mindset, and a strong platform engineering team, we can provide a foundation and integration layer to harmonize and empower software development and execution. Our challenge is to find the balance between the stability and security of the environments provided by us and an innovative drive that keeps evolving our platforms, ensuring that they age but don't retire.

But it is down to us. As with many things in life, if we don't believe in it and make it happen, no one else will. The platform approach is not the only way to go, yet it has a strongly growing group of experts, motivated contributors, and a lively community. We can observe at every major event how end users tell their success stories of introducing platforms and platform teams, and how it has become a game changer for their organizations.

As we look ahead, platforms stand at the forefront of our technological future, promising to simplify complexities and foster innovation. By focusing on platform engineering, architects can craft systems that are not only robust but also adaptable to the fast-changing landscape of technologies. Integrating platforms comes along with continuous learning and improvement, where small, thoughtful iterations can lead to significant improvements. We have to encourage a culture of collaboration and shared growth. While the challenges are many, the potential rewards of well-engineered platforms are indisputable, offering, as shown multiple times, a solid foundation to build resilient and scalable systems. Through careful planning and execution, platform engineering can unlock new possibilities for more efficient and effective systems. With a commitment and focus on our principles and purpose, we can create purely desirable environments that are future-proof and ready for change.

Summary

In this chapter, we discuss the necessity for change and how we, as platform engineers, have to actively embrace it. Through iterative evolutionary steps, we guide a platform through its lifetime, not by dismantling it but by maturing it. We keep it future-proof and provide innovative solutions. We also highlighted the importance of enabling sustainable approaches and architectures and emphasized that, for platforms, it is easy to provide optimizations centrally for the entire organization.

We discussed the golden path, its feedback loops, and a different perspective of it as a feature, not a purely opinionated setup. We emphasized the growth your platform will experience and the need to define golden paths to help it grow with new capabilities. We concluded the discussion by exploring the use of an existing golden path to create new golden paths.

Finally, we highlighted current trends and topics that might and might not shape the following years of our cloud-native universe. Who knows when the next big bang will happen and where it will lead us in the future?

This book provided you with many different perspectives, tools, and approaches to use for designing your own platform architecture. As an engineer or decision maker, you learned about the actual complexity of building platforms, which goes far beyond pure technology, and you learned about the value and benefits for users and organizations. As a platform architect, these steps and considerations support you along your journey, assisting you in creating an evolving product and a positive influence on those who implement and utilize it.

Going forward, it is down to you how you utilize what you have learned. If you are at the stage of defining your reference and target architecture, return to the **Miro** or **draw.io** templates *[5]*. We also would recommend joining the different platform engineering communities out there, such as the CNCF Platforms Working Group *[6]*, who frequently have knowledge-sharing sessions, and other users who provide insights into their approaches.

We wish you all the best for your journey and a lot of success with your platform.

Further reading

- [1] Spotify blog on Golden Paths: https://engineering.atspotify.com/2020/08/how-we-use-golden-paths-to-solve-fragmentation-in-our-software-ecosystem/

- [2] Backstage template documentation: https://backstage.io/docs/features/software-templates/writing-templates

- [3] Kratix: https://www.kratix.io/

- [4] Security incidents with virtual machines where researchers and shady people have broken out of the VM context:

  - VENOM Vulnerability (2015): The **VENOM (Virtualized Environment Neglected Operations Manipulation)** vulnerability was discovered in the QEMU VM's floppy disk controller. It allowed an attacker to escape from the guest VM and execute arbitrary code on the host machine: https://en.wikipedia.org/wiki/VENOM

  - CVE-2017-5715 (Spectre) and CVE-2017-5754 (Meltdown): These vulnerabilities exploited flaws in CPU design, affecting almost all modern processors. They allowed attackers to read sensitive data across VM boundaries by exploiting speculative execution: https://meltdownattack.com/

  - CVE-2019-11135: Known as ZombieLoad, this vulnerability in Intel CPUs can allow an attacker to leak data from other processes or VMs running on the same physical CPU: https://nvd.nist.gov/vuln/detail/CVE-2019-11135

  - Blue Pill (2006): Joanna Rutkowska demonstrated the Blue Pill attack, which showed how a hypervisor-based rootkit can be installed on the fly, essentially creating an undetectable VM escape: https://en.wikipedia.org/wiki/Blue_Pill_(software)

  - Cloudburst (2009): This was a VMware vulnerability that allowed code execution on the host machine from within a guest VM: https://nvd.nist.gov/vuln/detail/CVE-2009-1244

- And some more security incidcents with VMs: `https://en.wikipedia.org/wiki/Virtual_machine_escape`

- [5] Create your own architecture workshop:

  - `Miro` template: `https://miro.com/miroverse/platform-architecture-workshop`

  - `Draw.io` template: `https://github.com/PacktPublishing/Platform-Engineering-for-Architects/tree/main/Chapter%2002`

- [6] CNCF Platforms Working Group: `https://tag-app-delivery.cncf.io/wgs/platforms/`

Stay Sharp in Cloud and DevOps — Join 44,000+ Subscribers of CloudPro

CloudPro is a weekly newsletter for cloud professionals who want to stay current on the fast-evolving world of cloud computing, DevOps, and infrastructure engineering.

Every issue delivers focused, high-signal content on topics like:

- AWS, GCP & multi-cloud architecture
- Containers, Kubernetes & orchestration
- **Infrastructure as Code (IaC)** with Terraform, Pulumi, etc.
- Platform engineering & automation workflows
- Observability, performance tuning, and reliability best practices

Whether you're a cloud engineer, SRE, DevOps practitioner, or platform lead, CloudPro helps you stay on top of what matters, without the noise.

Scan the QR code to join for free and get weekly insights straight to your inbox:

https://packt.link/cloudpro

Index

`packtpub.com`

Subscribe to our online digital library for full access to over 7,000 books and videos, as well as industry leading tools to help you plan your personal development and advance your career. For more information, please visit our website.

Why subscribe?

- Spend less time learning and more time coding with practical eBooks and Videos from over 4,000 industry professionals
- Improve your learning with Skill Plans built especially for you
- Get a free eBook or video every month
- Fully searchable for easy access to vital information
- Copy and paste, print, and bookmark content

Did you know that Packt offers eBook versions of every book published, with PDF and ePub files available? You can upgrade to the eBook version at `packtpub.com` and as a print book customer, you are entitled to a discount on the eBook copy. Get in touch with us at `customercare@packtpub.com` for more details.

At `www.packtpub.com`, you can also read a collection of free technical articles, sign up for a range of free newsletters, and receive exclusive discounts and offers on Packt books and eBooks.

Other Books You May Enjoy

If you enjoyed this book, you may be interested in these other books by Packt:

Modern DevOps Practices

Gaurav Agarwal

ISBN: 978-1-80512-182-4

- Explore modern DevOps practices with Git and GitOps
- Master container fundamentals with Docker and Kubernetes
- Become well versed in AWS ECS, Google Cloud Run, and Knative
- Discover how to efficiently build and manage secure Docker images
- Understand continuous integration with Jenkins on Kubernetes and GitHub Actions
- Get to grips with using Argo CD for continuous deployment and delivery
- Manage immutable infrastructure on the cloud with Packer, Terraform, and Ansible
- Operate container applications in production using Istio and learn about AI in DevOps

Embracing DevOps Release Management

Joel Kruger

ISBN: 978-1-83546-185-3

- Discover the significance and anatomy of the SDLC

- Understand the history of release management and how various models work

- Grasp DevOps release management and basic strategies to implement it

- Construct optimized CI/CD pipelines capable of early issue detection

- Implement the shift-left approach to enhance value delivery to customers at record speed

- Foster a culture of cross-functional collaboration in your team

- Make DevOps release management pragmatic and accessible

- Overcome common pitfalls in DevOps release management

Packt is searching for authors like you

If you're interested in becoming an author for Packt, please visit `authors.packtpub.com` and apply today. We have worked with thousands of developers and tech professionals, just like you, to help them share their insight with the global tech community. You can make a general application, apply for a specific hot topic that we are recruiting an author for, or submit your own idea.

Share Your Thoughts

Now you've finished *Platform Engineering for Architects*, we'd love to hear your thoughts! Scan the QR code below to go straight to the Amazon review page for this book and share your feedback or leave a review on the site that you purchased it from.

https://packt.link/r/1836203594

Your review is important to us and the tech community and will help us make sure we're delivering excellent quality content.

www.ingramcontent.com/pod-product-compliance
Lightning Source LLC
LaVergne TN
LVHW080112070326
832902LV00015B/2535